PLATE I. Wild-type Drosophila melanogaster. Male at left, female at right. (E. M. Wallace, pinx.)

An Introduction
to
GENETICS

A. H. STURTEVANT
*Professor of Genetics, California Institute of
Technology
(now Thomas Hunt Morgan Professor of Biology,
Emeritus, California Institute of Technology)*

G. W. BEADLE
*Professor of Genetics, Stanford University
(now President, University of Chicago)*

DOVER PUBLICATIONS, INC.
NEW YORK

This Dover edition, first published in 1962, is a cor-
rected republication of the work originally published
by W. B. Saunders Company in 1939. This is an
unabridged republication of the first edition, except
that one color plate showing the eye-colors of *Drosophila melanogaster* has been eliminated.

Standard Book Number: 486-60306-7
Library of Congress Catalog Card Number: 62-53415

Manufactured in the United States of America
Dover Publications, Inc.
180 Varick Street
New York, N.Y. 10014

PREFACE

to Dover Edition

THIS BOOK represents the way genetics looked to us in 1939. In the past twenty-two years there have been far-reaching changes in the subject, but these have not been incorporated. The only changes that have been made are a few corrections of misprints and minor changes in wording, and the elimination of the plate showing the eye colors of Drosophila.

The following are some of the changes that a thorough revision would need to incorporate:

Chapter I. The chromosome number in man is now known to be 46, not 48.

Chapter VI. We should now place less emphasis on chiasmata, and would avoid complete equating of chiasmata and exchanges.

Chapter XII. This chapter needs revision, with discussions of complementation and fine-structure analysis.

Chapter XIII. The discussion of the mechanism of the induction of mutations by irradiation needs revision, and the chemical induction of mutations would play a large part in an up-to-date account.

Chapter XVI. Recent results indicate that the Y chromosome is much more important in sex-determination in mammals than it is in Drosophila.

Chapter XXII. The field of biochemical genetics is largely a new development, and this chapter needs expansion and revision.

There are some areas of genetics, now in the forefront of research on the subject, that are too recent to have been mentioned in the book. Any current account would be concerned with DNA (deoxyribose nucleic acid) as the probable chemical structure of the genes, and with the powerful new genetic techniques that have been developed with bacteria and bacteriophages.

In spite of these developments, however, we feel that the book gives a fair and balanced account of most of the field covered, i.e., of the chromosome mechanics of higher organisms. And this is an important, basic part of genetics that is often neglected or inadequately covered in recent books.

A. H. STURTEVANT

1962   G. W. BEADLE

PREFACE

GENETICS is a quantitative subject. It deals with ratios, with measurements, and with the geometrical relationships of chromosomes. Unlike most sciences that are based largely on mathematical techniques, it makes use of its own system of units. Physics, chemistry, astronomy, and physiology all deal with atoms, molecules, electrons, centimeters, seconds, grams—their measuring systems are all reducible to these common units. Genetics has none of these as a recognizable component in its fundamental units, yet it is a mathematically formulated subject that is logically complete and self-contained.

We have attempted to treat the subject in the way suggested by these considerations—namely, as a logical development in which each step depends on the preceding ones. The book should be read from the beginning, like a textbook of mathematics or physics, rather than in an arbitrarily chosen order, like a textbook of comparative anatomy or natural history.

Genetics also resembles other mathematically developed subjects, in that facility in the use and understanding of its principles comes only from using them. The problems at the end of each chapter are designed to give this practice. The student will find that it is important that they be actually solved.

This book is planned for the use of students who have had an introductory course in biology. Brief accounts of mitosis, fertilization, the life-history of a seed-plant, and similar topics, are given rather as reviews than as attempts to make the material intelligible to a student with no biological background. The treatment of the material is not a historical one; the object has been rather to give a natural order that would simplify the presentation. The last chapter has been added to give a picture of the order in which the essential ideas of the subject developed.

Crossing over is here presented from the first in terms of split chromosomes ("four-strand-crossing-over"). It is usual to give the unsplit-chromosome interpretation first, as being simpler and more easily grasped. This seems to us inexcusable, since it gives the student what is known to be an incorrect idea, thus making it more difficult than ever to grasp the correct one. The split-chromosome condition is not inherently difficult to understand, if one is given good diagrams and will learn to visualize them in three dimensions. The diagrams given here may look formidable; they cannot be understood at a glance, but we hope that study of them will convince the reader that the principles concerned are really simple. An understanding of these principles should make it possible to deduce the behavior of particular types of chromosomes, without memorizing individual schemes.

It may seem that the importance of chromosome aberrations has been overemphasized. The space devoted to them does not seem to us to be out of proportion to their importance. They have contributed largely to the clarification of general ideas in genetics, and constitute essential working tools in many of the attempts now being made to solve fundamental problems. The reader will find that many of these more general questions are discussed in the chapters whose titles might suggest that they deal only with special types of chromosome aberrations.

Since the book is intended as an elementary text, it has seemed to us that an extensive bibliography would be out of place. However, we have introduced a few references, chosen largely because they either contain extensive citations of the literature or are useful summaries. Their use will enable the student to find the important papers dealing with specific topics.

During the preparation of this book the authors have had the encouragement and helpful advice of many colleagues and friends. A number of persons have read the manuscript or parts of it. Doctors Barbara McClintock, Karl Sax, Berwind P. Kaufmann, D. F. Jones, and Miss Margaret Hoover have kindly supplied original prints of photographs. Miss Eugenia A. Scott has made a number of the original black and white drawings and diagrams.

Professor T. H. Morgan has kindly permitted the use of a number of figures prepared by Miss Edith M. Wallace, among which are the frontispiece and the plate of Drosophila eye colors. To all these persons the authors owe their gratitude.

A. H. STURTEVANT.
G. W. BEADLE.

CONTENTS

An Introduction to Genetics

CHAPTER I

SEX CHROMOSOMES

THE existence of diversity among organisms is one of the most familiar of natural phenomena. Every child recognizes not only the differences between dogs, cats, and men, but also those between different individuals of each of these species. It is obvious that these latter, individual differences, can be analyzed in part into separate components such as height, eye-color, or hair-form. Such an analysis leads easily to a classification, based on resemblances and differences between individuals—an obvious example being that of breeds of domestic animals or varieties of cultivated plants.

Variation.—*Genetics* is the science that deals with the underlying causes of these resemblances and differences. This is only another way of saying that it is the science of heredity and variation, for it needs no argument to establish the point that breed differences or such characters as skin-color in man are due to heredity. In practice the study of genetics is based largely on those characters with respect to which individuals fall into classes that are easily distinguishable, so that it is possible to make unambiguous classifications. Other kinds of characters can be analyzed, and will be discussed later in this book; but the cases of *discontinuous* variation are most easily dealt with.

Sex Differences.—The most widespread and generally recognized discontinuous character is that of sex. Most higher animals and many plants are represented by two types of individuals, male and female, with intermediates absent or very rare. Furthermore, in many kinds of animals and plants, the two sexes are present in approximately equal numbers. This relation is significant to the geneticist, for he is constantly studying the relative numbers of different kinds of individuals in families; and this simple 1:1 ratio at once suggests the presence of a simple mechanism.

Inheritance of Sex Differences.—The mechanism at work has been found, and is in fact simple. Cytologists have investigated the dividing cells of many kinds of organisms, and have found them to contain bodies, called *chromosomes,* that can be stained by certain techniques. As a rule the number of chromosomes is the same in all the body-cells of an individual, and in all the different individuals of a species. Not only is the number

Fig. 1.—Human chromosomes. At left, spermatogonial division. The Y (labelled) is the smallest chromosome, the X (unlabelled) is one of the three largest, the other two constituting a pair of autosomes. At right, side view of meiotic division of male, with X passing to lower pole, Y to upper, the autosomes being at the equator. (From Painter.)

constant, but so also are the relative sizes and shapes; furthermore, there are two of each kind. Thus, in man, the number in the cells of the body is 48 (Fig. 1); this is made up of single pairs of each of 24 distinct kinds, and it can be shown that one member of each pair is contributed by each parent.

There is one striking exception to the rule that the two members of a pair are alike. In many organisms, including man (Fig. 1), there is one pair whose members are visibly different in the male. In these cases the female has no such *heteromorphic* pair, both representatives of the pair in question resembling one of the two present in the male. The kind present in duplicate in the female is known as X, its dissimilar mate in the male as Y.

The formula for the female thus is XX, for the male XY. It must be remembered that this formula applies only to the one pair of chromosomes. For all other pairs the two sexes are alike.

Sex-Chromosomes.—The mature germ-cells or *gametes* (eggs and sperm) have half the number of chromosomes present in the body-cells of the individual that produces them—one member of each pair. This result is brought about by the *meiotic* divisions (in animals the last two divisions before the production of the

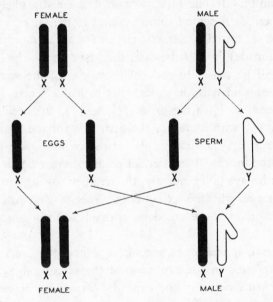

Fig. 2.—Schematic representation of the distribution of X and Y chromosomes.

gametes), which will be described in more detail in later chapters. For our present purposes it need only be emphasized that each gamete receives one chromosome of each pair present in the body-cells. It follows that, for the pair in which the male and female differ, all the eggs carry one type (X) while the sperm are of two types (X and Y) in equal numbers. As the accompanying diagram (Fig. 2) indicates, random fertilization results in an average ratio of 1 XX (female) to 1 XY (male) in the next generation.

Sex-ratio.—The 1:1 sex-ratio is an average; it is a matter of everyday experience that it is not to be expected in every family. The relations here are similar to those encountered in the tossing of coins, and they are those dealt with in the theory of probability, an understanding of which is necessary in genetics, since the whole subject is based on the study of ratios of classes; and every family and every experimentally obtained ratio is subject to what is known as the error of sampling.

Probability.—If one tosses a coin it is equally likely to fall heads or tails. A second coin is subject to the same probability. If the two are tossed in succession, then (on an average, if enough tosses are made) in half the cases the first coin will be heads, in half it will be tails. In each of these events, the second coin will be heads in half the cases, tails in half. There will thus be two heads in $\frac{1}{4}$ of the tosses ($\frac{1}{2} \times \frac{1}{2}$), one head and one tail in $\frac{1}{2}$ ($\frac{1}{4}$ with heads for the first, tails for the second, plus $\frac{1}{4}$ with tails for the first, heads for the second), and two tails in $\frac{1}{4}$ of the tosses. The essential point to remember is that the two coins behave independently; the result of tossing the first has no influence on what the second does. They may be tossed simultaneously, or on successive days, without any difference in the result.

In general, if one tosses n coins at a time, the average result can be expressed by the expansion of the expression $(a + b)^n$, where the coefficients in the expanded expression represent the number of cases found, the exponent of a the number of heads in the toss, and the exponent of b the number of tails. For example, for three coins the result is $a^3 + 3a^2b + 3ab^2 + b^3$; *i.e.,*

three heads	1 time
two heads, one tail	3 times
one head, two tails	3 times
three tails	1 time
Total	8 ($= 2^n$)

In terms of families, these proportions become probabilities. Thus, for example, one may deduce that a family of three children has one chance in 8 of including three boys. A similar argument will show that, on the average, one in 16 families of four will include only girls. A further consideration of the relation of probability to genetics may be found in the Appendix.

The X-Y mechanism is, in fact, the cause of the approximate 1:1 sex-ratio in man and many other organisms. It is subject to many modifications, but everywhere the basis of the 1:1 ratio remains the segregation of unlike members of a pair of chromosomes. Commonly, as in man, X is larger than Y; in some cases (*e.g.,* Drosophila melanogaster) X is smaller than Y; in other cases (*e.g.,* some bugs) X and Y are alike in size and shape; in still other cases (*e.g.,* grasshoppers) Y is absent. In some cases (birds, moths) the heteromorphic pair occurs in the female instead of in the male. This and still other types of sex-determination (some of which do not result in a 1:1 ratio) will be discussed in Chapter XVI.

It should be pointed out that the ratio of 1:1 is only approximately realized. There are real deviations in many species. In man the ratio, at birth, is about 106 males to 100 females; since stillbirths are more often male than female the discrepancy is presumably still greater at the time of fertilization, which is the stage at which the 1:1 ratio is expected according to theory. The reason for this deviation is not clear, though the most likely view seems to be that the X and Y sperm swim at slightly different rates, so that a Y sperm is more likely to reach the egg first. In any case, there can be no doubt that the two kinds of sperm are produced in equal numbers.

Theories of Sex-Determination.—Literally hundreds of theories have been proposed to account for the determination of sex, their construction having been for centuries a favorite occupation of philosophers and theoretical biologists. With the discovery of the sex-chromosome mechanism, by McClung, Stevens, and Wilson (1901–1905), the whole character of the problem was changed. The old speculative theories were forgotten, and the new ones that are still being proposed have no scientific standing.

This does not mean that there is nothing more to be done in the field; like every important scientific discovery, the sex-chromosome mechanism gives the solutions to many problems, but at the same time makes it possible to investigate still other questions that were either not apparent or not approachable before. In Chapter XVI we shall consider some of the more recent studies on the nature of sex-determination.

REFERENCES

(See also references for Chapter XVI.)

Schrader, F. 1928. The sex chromosomes. 194 pp. Gebr. Borntraeger, Berlin.

Wilson, E. B. 1925. The cell in development and heredity. 1,232 pp. The Macmillan Co., New York.

PROBLEMS

(Note: In problems involving probabilities see Appendix for methods. Assume that the sex-ratio in man is 1:1.)

1. In a family of five children what is the probability that:
    (a) All children will be girls?
    (b) All children will be of one sex?
    (c) There will be three boys and two girls?
    (d) The family will not be made up of boys only?
    (e) There will not be more than one boy, *i.e.,* there will be one boy or none?
2. If the first child of a family is a girl, what is the probability of the second child also being a girl? If the first five children are girls, what is the probability that the sixth will also be a girl?
3. Considering two separate families, A and B, of three children each, what is the probability that family A will contain only boys and family B only girls? What is the probability that one or the other of the families will contain only boys and the remaining one only girls? What is the probability that the two families together will consist of three boys and three girls without regard to their distribution within the families?

CHAPTER II

SEX-LINKAGE

Drosophila.—Much of the modern theory of genetics is based on studies of a small fly called Drosophila melanogaster (see *Plate* I). This animal can be bred in the laboratory in large numbers with relatively little trouble and expense. A sexual generation, from egg to egg, may be completed in ten days; a single pair will produce hundreds of offspring. This species has a relatively small number of chromosomes—only four pairs

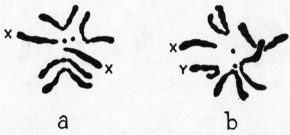

Fig. 3.—Chromosomes of Drosophila melanogaster. a, female; b, male. (After Dobzhansky, from Morgan, "The Scientific Basis of Evolution," W. W. Norton & Company, Inc., Publishers.)

(Fig. 3). These are some of the characteristics that have led to the widespread use of this species.

Drosophila melanogaster is to be found in all temperate and tropical countries, and all wild strains, wherever found, are essentially alike. This relatively uniform type, to be found about fermenting fruit in most parts of the world, is technically known as the *wild type*. Occasional variants do occur, both in nature and in the laboratory. Many of these transmit their peculiarities to their descendants, and may be isolated in true-breeding strains. During the past thirty years hundreds of such strains have been accumulated, and are now maintained in many laboratories. They constitute the working material of the geneticists who study this species.

Mutant Characters.—Each new variation that is inherited must evidently arise through some change in the material basis of heredity. Such a change is called a *mutation,* and the new character due to it is a *mutant character.* Many such types will be discussed in this book.

One of the mutant characters, known as bar, consists in a reduction of each compound eye to a narrow vertical bar, in contrast to the round of the normal or wild-type (Fig. 4). A bar strain breeds true to the character. If a wild-type female is mated to a bar male, the offspring consist of wild-type sons, and of daughters with eyes somewhat like those of the bar father, but

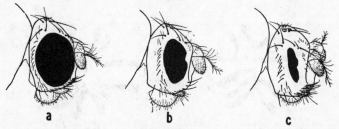

a   b   c

Fig. 4.—Eyes of Drosophila melanogaster. a, wild-type; b, wide bar female; c, narrow bar.

definitely broader—*i.e.,* intermediate between wild-type and the narrow bar of the pure strain, and distinguishable from both (Fig. 5—for the time being, disregard the lower four flies in this figure). When the reciprocal mating is made, using a female from the narrow bar stock and a wild-type male, the offspring are broad bar daughters (like those from the first mating) and narrow bar sons, like those from the bar stock (Fig. 6—disregard the lower four flies in this figure for the present).

Bar and the Sex-Chromosomes.—These relations suggest those which hold for the X chromosomes of this species. Every male gets his single X from his mother; in both of the crosses described, the males have eyes indistinguishable from those present in the strains from which their mothers come. Every female gets an X from each parent; in both crosses the females have eyes intermediate between those of their two parents. The hypothesis is suggested, that the difference between wild-type and bar is due

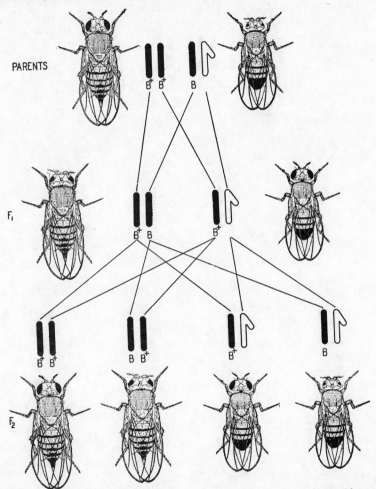

Fig. 5.—Diagrammatic representation of the results of crossing a wild-type female to a bar male.

to a difference between their X chromosomes—that the mutation from wild-type to bar was caused by some sort of change in an X chromosome.

One test of this hypothesis may be made by breeding from the flies obtained in the experiments already described. The immediate offspring, discussed above, are said to belong to the "first filial

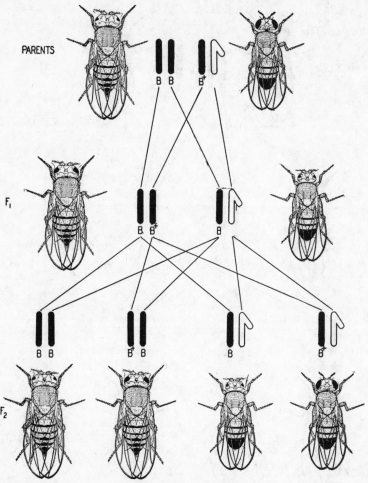

Fig. 6.—Diagrammatic representation of the results of crossing a narrow bar female to a wild-type male.

generation," regularly abbreviated to F_1. If the broad bar females from the first cross are mated to their wild-type brothers, their offspring are said to constitute the F_2 generation. In the present case these consist of broad bar females, wild-type females, narrow bar males, and wild-type males—each of these occurring as one-fourth of the total.

The result is in agreement with the hypothesis, as figure 5 shows. The element in the X chromosome that is responsible for the bar eye is the bar *gene.* This will be used in complex formulae; accordingly it is useful to have a simple symbol for it. For this purpose the name of the mutant character is abbreviated, the symbol of the gene being "B." The mate to this gene, present in the wild-type X, is designated "B^+" (sometimes written "$+^B$"). The plus sign is used in general to indicate the genes present in the wild-type fly. The broad bar F_1 female produces eggs, half of which carry a wild-type X, the other half a bar one. The F_1 males produce sperm, half of which carry a wild-type X, the other half a Y. As figure 5 shows, random fertilization of these gives the observed result: one female with two B^+ X's (wild-type); one female with one B and one B^+ (broad bar); one male with a B^+ X (wild-type); one male with a B X (bar).

Segregation.—One further deduction appears from this analysis: the B^+ and B genes, each present in an X of the F_1 female, are recovered in their original form. The types obtained in F_2 are not in any way different from the corresponding types present in the parental strains or in the F_1 generation. This *segregation* is one of the fundamental principles of genetics.

The production of B and B^+ eggs in equal numbers is not quite as simple a matter as is the production of X and Y sperm in equal numbers. In the male, there are *diploid* cells—*i.e.,* cells having the full double set of chromosomes, one pair of each kind —in the testes, called spermatocytes, which undergo the two successive meiotic divisions, to give rise to what are known as spermatids (Fig. 7). The latter are *haploid*—*i.e.,* each has a single set of chromosomes, one representative of each kind. (The Greek root of the word haploid means "single.") The spermatids develop directly into functional sperms, without further nuclear divisions. It follows, therefore, that each diploid spermatocyte gives rise (by means of the two divisions mentioned) to four haploid sperms, of which two carry one member of any given pair of chromosomes and the remaining two carry the other member. Since all are potentially functional, the mechanism insures the production of exactly equal numbers of sperm carrying each member of the original pair of chromosomes.

In the female the last diploid cell in the cycle is the oöcyte. The two meiotic divisions result in four haploid cells (Fig. 7); but only one of these, the egg, takes part in development. The

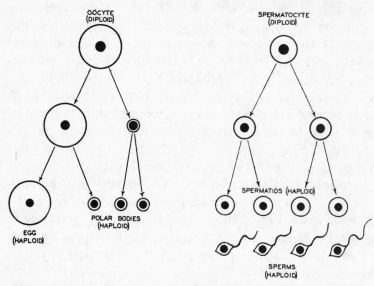

Fig. 7.—The formation of the gametes in animals. The two meiotic divisions result in one egg and three polar bodies (female, at left), or four sperms (male, at right).

other three nuclei, called polar bodies, later degenerate. Thus each diploid oöcyte gives rise to a single haploid egg. One member of each pair of chromosomes is present in this egg, but the other member of the pair is lost, in two of the polar bodies. However, it is a matter of chance which member of the pair is present in the egg; accordingly, a series of eggs produced by a single individual will be constituted as are the sperm produced by a male, i.e., half will carry one member of the original pair, half the other member.

Prediction of Results.—Further tests of the interpretation given above for the case of bar may easily be made. In general, the offspring of any mating involving B may be predicted. The F_2 from the reciprocal mating (bar female by wild-type male) may be taken as an example (Fig. 6). The F_1 broad bar female gives eggs with a B^+ X or with a B X. These will occur in equal numbers. The F_1 male gives sperm with a B X or with a Y, again in equal numbers. There result, in F_2: one narrow bar female, one broad bar female, one bar male, one wild-type male. It will be noted that the breeding behavior of any individual may be predicted by observing the shape of its eyes. A narrow bar female has two B X's which is written $\dfrac{B}{B}$, with each horizontal line standing for an X chromosome. This is often simplified to $\dfrac{B}{B}$, where the second horizontal line is omitted; or, to save space and trouble in typing or printing, B/B. A broad bar female is B/B^+, a bar male B/Y, a wild-type male B^+/Y. These constitutions specify the properties of an individual, quite apart from its pedigree; one male that is B/Y is of exactly the same nature as another with the same formula, though one may have come from a pure strain whereas the other may be from an F_1, half of whose recent ancestors were wild-type.

Genes and Chromosomes.—The X chromosome, then, is not solely concerned with the determination of sex. It also carries the units, called genes, that we have been discussing under the names of B^+ and B. There are many mutant types that are inherited in the same way, so that many genes must be contained in the X chromosomes. This type of inheritance is called *sex-linkage,* and the genes concerned are said to be *sex-linked*. The case of bar is somewhat unusual in one respect. In most cases it is not possible to distinguish easily, by inspection alone, individuals with one $+$ gene and one mutant gene from those with two $+$ genes. A typical example of this sort may now be considered.

Dominance.—One of the mutant strains has pure white compound eyes, the color in the wild-type being a deep red. If a white-eyed female is mated to a wild-type male, the F_1 offspring are red-eyed (wild-type) females and white-eyed males (Fig. 8). The reciprocal mating (wild-type female by white male) gives wild-type females and wild-type males in F_1 (Fig. 9). Here again the F_1 males resemble the maternal stock, and the F_1 females from the two crosses are alike. But the F_1 females are also, in appearance, like the wild-type strain, rather than inter-

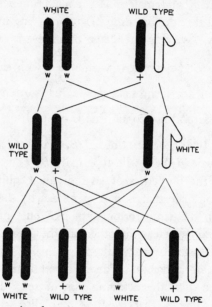

Fig. 8.—The results of crossing a white-eyed female to a wild-type male.

mediate as in the case of bar. This is described by saying that the w^+ gene is *dominant,* the white gene (w) is *recessive.* The diagrams (Figs. 8 and 9) show that the F_2 results confirm the interpretation. The F_1 females give w^+ eggs and w eggs in equal numbers, and the results are all in agreement with those from bar, except only that B^+ and B show no dominance, but give an intermediate.

Terminology.—The terminology for the genes used here needs an explanation. The symbol "$+$" has in fact been used to represent two different things; the B^+ that acts as a mate to B and is concerned with eye-shape is not the same thing as the w^+ that acts as the mate of w and is concerned with eye-color. If one wished to be exact, they should always be distinguished as B^+ and w^+, as has been done here, but the simplified form $+$ —with no base—is usual in cases where confusion does not arise.

Certain technical terms, to be used later, can now be introduced. The genes B and B^+ are spoken of as *alleles* (or allelomorphs),

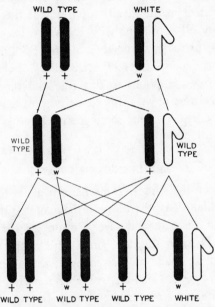

Fig. 9.—The results of crossing a wild-type female to a white-eyed male.

and the relation between them is that of *allelism*. The essential points in this relationship are that they act as mates at segregation, and that they are concerned in the development of contrasted characters of the organism. An individual with two like alleles (*e.g.*, w/w or $+/+$) is said to be *homozygous*, one that has two unlike ones (*e.g.*, $B/+$) is said to be *heterozygous*. The relation of dominance may be reworded thus: when a heterozygote is not distinguishable (without breeding tests) from one homozygote, then the allele present in that homozygote is said to be dominant to the one present in the distinguishable homozygote. This relation is only a relative one. We have seen that, in the case of bar, dominance in this sense is absent. Strictly speaking, it may be questioned whether $w/+$ is really identical in appearance with $+/+$; certainly under special conditions it may be made distinct. This relative nature of dominance will be discussed again.

Y Chromosomes.—It will be observed that we have not taken into account the Y chromosome, which is the mate of the X, and segregates from it in the meiotic divisions of the male. The results given are all consistent with the view that it may be neglected. This turns out to be the rule for sex-linked genes; the Y does not, in general, carry effective alleles of sex-linked genes. There are a few exceptions to this rule, but they are rare, and need not concern us now. This result is consistent with the fact, mentioned in Chapter I, that some species (such as grasshoppers) have no Y at all.

SEX-LINKAGE IN MAN

We have seen that the X-Y mechanism exists in man, as well as in Drosophila. It is, therefore, to be expected that sex-linkage will also occur in man, and several examples are known. The analysis here is more difficult, since pure strains do not exist, matings cannot be made as desired by investigators, families are small, and few generations are available for examination at any one time. The subject must be studied through the accumulation and analysis of pedigrees showing the inheritance of specified characters.

Color-blindness.—Perhaps the best-known example of sex-linkage in man is that of color-blindness. There are several types of color-blindness known; but at least the common type of red-green blindness behaves as a sex-linked recessive (see figure 10 for a pedigree). Color-blind men are found much more frequently than are color-blind women, since the latter must receive the mutant gene from both parents—and in fact color-blind women do have color-blind fathers—whereas a heterozygous

woman, herself with normal vision, will have sons half of whom
are color-blind, regardless of the constitution of their father.

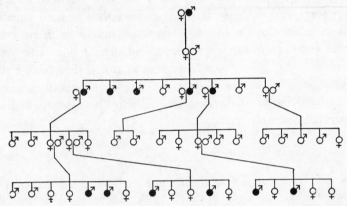

Fig. 10.—A human pedigree illustrating the inheritance of red-green color-
blindness. The black circles represent affected individuals. ♀ signifies female,
♂ signifies male. (After Holmgren and Göthlin.)

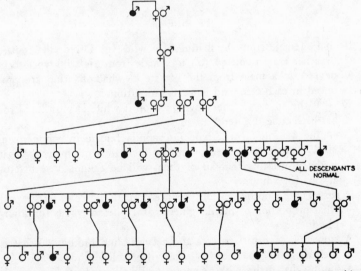

Fig. 11.—Human pedigree illustrating the inheritance of hemophilia. The black
circles represent "bleeders." (From Mohr, after Schloessmann.)

Hemophilia.—Another sex-linked recessive gene in man is
that for hemophilia, or "bleeding" (Fig. 11). In this type the

blood fails to clot easily on exposure to air, so that even minor wounds are often fatal. A normal woman heterozygous for hemophilia produces sons half of whom are bleeders. These rarely survive to the reproductive age, since the disease is usually fatal in infancy or childhood. The result should be that homozygous females are rarely produced, and in fact none has been recorded. This gene has been present in certain European royal families, and, because the eldest sons of the last Czar of Russia and of the last King of Spain were afflicted with the disease, has had certain political consequences.

REFERENCES

Davenport, C. B. 1930. Sex linkage in man. Genetics, 15: 401–444.

Morgan, T. H., and C. B. Bridges. 1916. Sex-linked inheritance in Drosophila. Carnegie Inst. Wash., publ., 237. 87 pp.

Morgan, T. H., C. B. Bridges, and A. H. Sturtevant. 1925. The Genetics of Drosophila. Bibliogr. Genetica, 2: 1–262.

PROBLEMS

1. If an F_1 female from the mating of a wild-type Drosophila female by a bar male is mated (a) to a male from the wild-type stock, or (b) to a male from the bar stock, what offspring are produced in each case, and in what proportions?

2. What is the answer to the above question if white is substituted for bar each time that term appears?

3. From the mating of wild-type Drosophila female by white male, six F_2 females are tested by mating them to white males. What is the probability that all of them will be found to be heterozygous for white?

4. If parents with normal vision produce normal and color-blind sons, what proportion of the daughters will be heterozygous for colorblindness?

5. A woman has normal parents and a color-blind brother. What is the probability that her first son will be color-blind?

6. Assuming that males with hemophilia never reproduce, from which one of the grandparents of an individual with the character did the gene for hemophilia come?

7. A normal woman whose father was color-blind marries a man with normal vision. What proportion of her sons are expected to be

color-blind? If her husband had been color-blind, what would
have been the expectation for the sons?

8. Look up in the Encyclopedia the pedigrees of the two princes referred to in the text. Can you trace the gene for hemophilia
   back to a common ancestor?

CHAPTER III

AUTOSOMAL INHERITANCE

Most of the chromosomes occur in the body-cells in pairs the members of which are alike. We have considered the special case of the X chromosome, which is represented only once in the male. The other chromosomes also contain genes, and the behavior of these chromosomes in both sexes is like that of the X's in the female. This is the more common type of inheritance, and the one that was first worked out by Mendel.

Mendel.—Gregor Mendel, a monk at Brünn, carried out crossing experiments with garden peas, which led to the formulation of the fundamental principles of genetics in terms that are still satisfactory. The paper describing and explaining his results was published in 1866, but made no impression on his contemporaries. It was, in fact, forgotten until after Mendel's death. In 1900 it was brought to light by Correns, de Vries, and Tschermak, all of whom had been carrying on crossing experiments independently, and were able both to confirm Mendel's conclusions and to recognize their significance. The modern science of genetics began with this triple confirmation.

Yellow and Green Peas.—Some of Mendel's strains of peas had yellow seeds, others had green. This color is in the cotyledons, which form part of the plant that is to develop from the seed, not of the plant that bears the seed. Accordingly, it is not surprising that the two reciprocal crosses (yellow ♀ × green ♂, green ♀ × yellow ♂) give the same result, in the seed resulting immediately from cross-pollination. This F_1 seed is yellow in both cases—*i.e.,* yellow is dominant. A number of different conventions are in use for the designation of the genes concerned here; Mendel arbitrarily called them *B* (yellow) and *b* (green), but we shall name them from the recessive allele, as though yellow is considered "wild-type." That is, the green parent is g/g, the yellow g^+g^+, or simply $+/+$, and the F_1 yellow is $+/g$.

The 3 to 1 Ratio.—Mendel's F_2 generation was obtained by *selfing* the plants grown from the F_1 yellow seeds; *i.e.*, by pollination of a plant with its own pollen—self-pollination. In peas both male and female organs are present in the same flower, and self-fertilization is the usual method of reproduction. It was, therefore, only necessary to raise the F_1 plants and harvest the

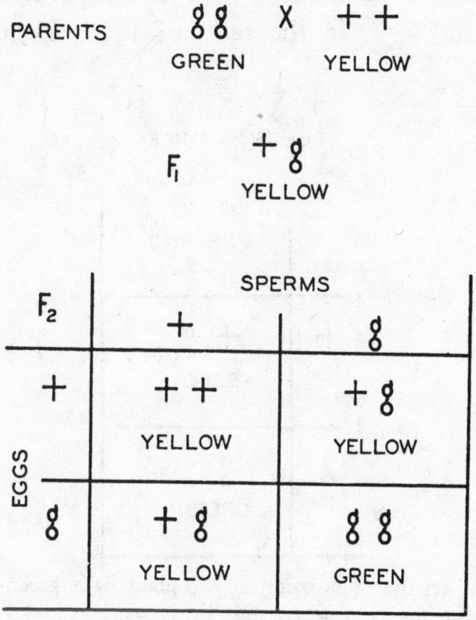

Fig. 12.—The results of a cross between green and yellow peas. The symbols above and to the left of the "checkerboard" indicate the gametes produced by the F_1 plants.

seed produced by them. The result was a count of 6,022 yellow seeds, 2,001 green ones.

As shown in figure 12, the principles of segregation and random fertilization would lead to the following results: half the eggs are $+$, half are g; the same proportions occur in the sperms. The F_2 should then be made up of $+/+$, $+/g$, g/g, in the proportions 1:2:1. Since the F_1 seeds were yellow in color, it follows that $+/g$ is yellow, like $+/+$. The expected ratio therefore is 3 yellow : 1 green. It is obvious at once that this is the ratio actually obtained. Given 8,023 seeds, the expected numbers are 6,017¼ yellow to 2,005¾ green—or, since the fractions are meaningless, 6,017 to 2,006. The observed result differs from this by 5 seeds; *i.e.*, if 5 of the yellows had been green the fit would have been as close as possible.

The 1:2:1 Ratio.—Mendel verified the conclusions as to the mechanism involved here, in several ways. The F_2 greens were shown to breed true, as expected of homozygotes (g/g). The F_2 yellows should be made up of homozygotes ($+/+$) and heterozygotes ($+/g$), in the ratio of 1:2. Yellow seeds were

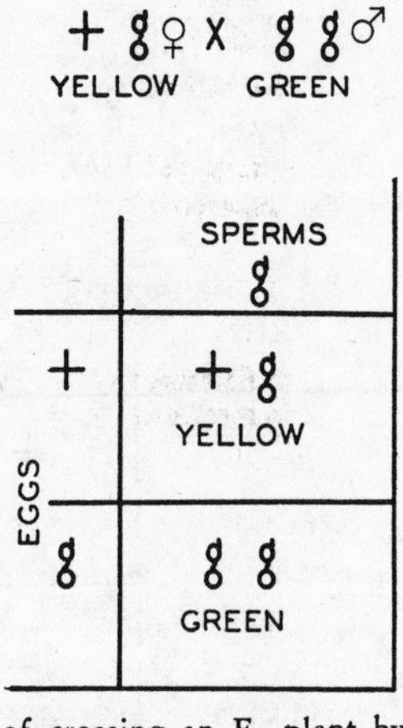

Fig. 13.—The results of crossing an F_1 plant by a green-seeded one. The cross is shown for the case where the F_1 is used as the female parent; the reciprocal mating (green pollinated by F_1) gives the same result.

sown, and the seeds (F_3, resulting from self-fertilization) borne by them were examined. Of the 519 plants so obtained, 166 bore only yellow seeds (and hence were $+/+$), 353 bore yellow and green in the ratio of 3:1 (and hence were $+/g$). The expected numbers, on the 1:2 basis, are 173 to 346—*i.e.*, the deviation is 7. Clearly, then, the true F_2 ratio was, as expected, 1:2:1; the observed 3:1 was due to the obscuring effect of the dominance already observed in F_1.

Testcrosses.—The simplest way to test the hypothesis is to mate the heterozygotes to the recessive (green) strain, for here no dominant genes are brought in by the homozygous recessive parent to obscure the result, and a simple 1:1 ratio may be expected to result from the production of + and g gametes in equal numbers by the heterozygous parent (Fig. 13). Mendel carried out this experiment, and found that the F_1 +/g actually does produce equal numbers of + and g eggs (tested by pollinating by g/g), and also equal numbers of + and g sperm (tested by pollinating a g/g plant by a heterozygote).

Chromosomal Basis.—The parallelism to the behavior of the members of an ordinary pair of chromosomes (called *autosomes,* in distinction to sex chromosomes) is clear. The body-cells are duplex, each containing a member of each pair of chromosomes from one parent and one member from the other parent; they also carry one allele of each pair of genes from one parent, the other member from the other parent. Meiosis results in segregation of the members of the pairs, so that each gamete contains only a single chromosome of each pair, and likewise only a single allele of each pair of genes. In technical terms, these relations are expressed by saying that the body-cells are *diploid,* the gametes are *haploid.*

Other Characters in Peas.—Mendel also found a series of other characters in his peas that were inherited in this same way. Some strains had wrinkled seeds, others had round ones. This again is a property of the cotyledons, so that the seeds belong to the generation of the plant they are to grow into. In this case the F_1 seeds were round, and F_2 included 5,474 round and 1,850 wrinkled seeds. Evidently round is dominant, and this is a reasonable approximation to a 3:1 ratio.

Other characters were studied on the plants themselves. A tall and a dwarf variety gave a tall F_1 generation; in F_2 there were 787 tall to 277 dwarf plants. On crossing a variety with colored flowers to one with white flowers, the F_1 was found to have colored flowers; in F_2 there were 705 colored and 224 white flowered plants. In these and other cases the approximation to the expected 3:1 ratio is satisfactory.

Incomplete Dominance.—Since Mendel's time it has been

shown that dominance is not complete in one of the cases he studied—namely that of round *vs.* wrinkled seeds. Microscopic examination shows that starch grains in the cotyledons of the two homozygous strains are different; those in the heterozygous seeds are intermediate, even though these heterozygotes look, to the naked eye, quite like the homozygous rounds.

In some other instances the heterozygous class is easily and sharply distinguishable from both homozygotes. One well-known case of this sort is that of the Blue Andalusian fowl. The feathers of this breed are marked with a very fine mosaic of black and white, the effect being a slate blue. Breeders have long known that it is not possible to "fix" this color; blues mated together regularly produce not only blue offspring, but also black ones, and a third type that is white, with scattered splashes of blue. The ratio between these three types is 1 black to 2 blue to 1 splashed white. This ratio at once suggests that we are dealing with a Mendelian case, where black and splashed white are the two homozygotes, and blue is the heterozygote. Tests show that this supposition is correct; blue mated by black gives 1 blue to 1 black; black by splashed white gives only blue offspring; blue by splashed white gives 1 blue to 1 splashed white; both black and splashed white, when each is mated to its like, breed true. Here, then, as in the case of bar discussed in Chapter II, the heterozygote is intermediate, and dominance is absent.

It was assumed, in describing the yellow-green pair of genes for seed color in peas, that dominance is complete—that $+/+$ and $+/g$ are really identical except in their breeding behavior. The question has been raised, whether really complete dominance ever occurs; however, for all practical purposes it is certain that dominance may be treated as complete in the majority of cases.

Dominance in Measurable Characters.—When one is dealing with quantitative characters (*i.e.,* characters that can be counted, measured, or otherwise determined in numerical terms) it is possible to express the degree of dominance in exact terms. In the case of the bar eye of Drosophila, each eye is made up of a number of separate elements (ommatidia, with superficial lenses or facets). These can be counted; the results of such counts are shown in figure 14.

An autosomal recessive in Drosophila, known as "stubbloid,' reduces the length of the bristles (Fig. 15). Here dominance is

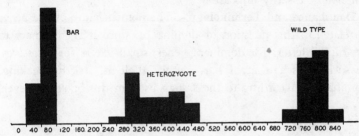

Fig. 14.—Facet numbers in Drosophila. The horizontal axis represents facet numbers; the vertical axis percentage frequencies for each of the three types shown. Each type is somewhat variable, but the three curves are wholly distinct.

practically complete, the heterozygote being indistinguishable from the homozygous wild-type. As shown in figure 16, measurements of the length of a particular bristle show that, while it is possible that the average bristle length is less in the heterozygous class, the distribution curves for it and for the homozygous wild-

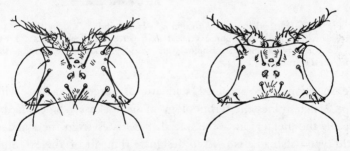

Fig. 15.—Heads of Drosophila. Wild-type to left, homozygous stubbloid to right. (After Dobzhansky.)

type are so nearly identical that classification of individual specimens is quite impossible.

The two examples illustrated by the curves represent extreme cases, that to which stubbloid belongs being the more usual type. There are, however, many instances of intermediate nature—

where the curve for the heterozygote is clearly different from that for either homozygote, but overlaps one or both of them either slightly or widely. In still other cases the curves for the two homozygous classes may overlap; analysis then becomes difficult but not necessarily impossible.

Dominance and Terminology.—The nomenclature of the genes is related to this question of dominance, since it is customary to use capital letters for dominant genes, small letters for recessives. In such cases as that of bar, neither allele is, strictly speaking, dominant. According to the system used in this book, however,

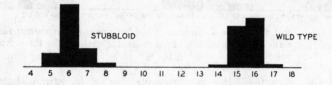

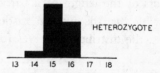

Fig. 16.—Bristle length in Drosophila. Horizontal axis, length of a particular bristle in arbitrary units; vertical axis, percentage frequencies for each of the three types shown. Homozygous wild-type cannot be distinguished from the heterozygotes.

the selection of a specified standard of reference (the "wild-type") determines the terminology in such a case. Since the bar type is the mutant one—the one that deviates from the standard wild-type—its name is used to designate the pair of alleles. Since the heterozygote, carrying one mutant gene and one + gene, is easily distinguishable from the wild-type, the symbol is capitalized, *B*. White is also a mutant type, and gives its name to the pair of alleles concerned; in this case the heterozygote cannot be distinguished from wild-type, so the small letter *w* is used. It should be emphasized that this system has been developed only as a matter of convenience; specifically, no theoretical importance is to be attached to the convention of treating bar as a dominant —the fact is, however, that in actual experiments it is most often convenient to plan to distinguish $B/+$ from $+/+$, rather than $B/+$ from B/B; and the former plan amounts to using bar as a dominant mutant type.

Taste Reaction.—Examples of Mendelian differences cited so far have been chiefly those in which we were concerned with visible differences—colors, sizes, shapes. These are usually the most convenient types to work with; but there are other, less obvious kinds of characters that show the same kind of inheritance. For example, in man it has been found that individuals differ in their ability to taste an organic chemical called phenylthiocarbamide (known also as phenyl thiourea). As ordinarily used in tests, this substance has a strong and unpleasant bitter taste to about 7 out of 10 persons, while to the remaining 3 it is tasteless. Analysis of pedigrees shows that this ability to taste the substance is due to a dominant gene. When both parents are non-tasters, all the children are also non-tasters. Families with one or both parents as tasters give proportions of the two classes in agreement with expectation (see Chapter XVIII for more detailed discussion).

CHARACTERS CLASSIFIABLE IN HAPLOID TISSUES

Waxy Maize.—There are a few instances in which segregating genes produce effects that may be identified in the haploid stage of the life-history; in such cases the segregation is very striking, since it is not necessary to deduce what has happened by taking into account the contribution from the other parent. Such an example is that of the waxy character in maize (Indian corn, or in American terminology, corn). This character, found in certain strains, was originally named from the waxy appearance of the kernels. The most certain method of identification is by staining with iodine. Most varieties give a blue color with this treatment —the familiar color obtained on treatment of starch with iodine. The starch in waxy kernels, however, gives a red color on such treatment. If normal and waxy strains are crossed, the F_1 is normal, and in F_2 a ratio of 3 normal to 1 waxy is obtained. In other words, the waxy character is due to a recessive gene (*wx*). The peculiar red color on staining with iodine also occurs in the starch present in some parts of the plant other than the kernels —*e.g.,* in the pollen grains. If the pollen grains of an F_1 plant (from the crossing of normal and waxy strains) are stained, it is at once evident that half of them are blue, half are red.

Waxy and the Life Cycle of Maize.—The significance of this result, and of others concerning the distribution of the red-staining starch in the hybrids, becomes clear if we consider the behavior of the chromosomes in the life-cycle of the maize plant (Fig. 17).[1] The diploid tissue, corresponding to the body-cells of an animal, is known as the sporophyte, and constitutes most of the plant ordinarily seen—the stem, leaves, roots, flowers. In the male organs (anthers) and in the female ones (ovaries) lie certain cells which undergo the two meiotic divisions and thereby produce haploid cells. These cells do not, as in animals, form the gametes directly, *i.e.,* without further divisions; on the contrary they undergo two divisions (in the anther) or three (in the ovary), to produce small structures that are haploid, and are known as the male and female gametophytes, respectively. One nucleus from a gametophyte functions as a gamete (sperm or egg), and the fusion of these at fertilization initiates the development of a new sporophyte (diploid, since each gamete introduces a haploid set of chromosomes).

Male Gametophyte.—The pollen grain is the male gametophyte. The cell from which a single pollen grain develops, which is haploid and the immediate result of two meiotic divisions, is known as the microspore; the cell that undergoes two meiotic divisions to produce four microspores is known as the microsporocyte (sometimes called the pollen-mother-cell, by some authors abbreviated to PMC). The nucleus of the microspore divides (without a division of the cell itself); one product is the tube-

---

[1] The account of the life history that follows applies to maize. It may be taken, however, as applying also to seed-plants in general, with the proviso that some of the details differ from one group of plants to another.

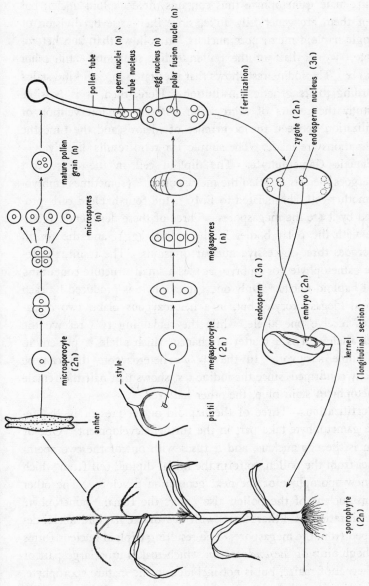

Fig. 17.—Diagram of the life history of the maize plant. Arrows represent the passage of time, lines without arrow-heads do not. Nuclei shown with broken lines degenerate, either at once or after a few divisions. The letter "n" signifies one complete (haploid) set of chromosomes.

nucleus, the other divides again to form two sperm-nuclei. The mature male gametophyte thus contains three haploid nuclei; but all of them are genetically alike, since they arise by division of a single haploid microspore nucleus. It follows that, in a hetero-zygote, $+/wx$, half of the pollen grains contain $+$, the other half wx. The iodine test shows that the starch in each kind stains according to its genetic constitution. Here, then, it is possible to study the results of segregation without the intervention of fertilization between the occurrence of meiosis and the time the observations are made. The simple 1:1 ratio results directly.

Female Gametophyte.—The diploid cell in the ovary that undergoes meiosis is called the megasporocyte (sometimes embryo-sac mother-cell, abbreviated to EMC), the four haploid cells pro-duced by it are the megaspores. Three of these degenerate (com-pare with the polar bodies in the animal egg), and the fourth undergoes three successive nuclear divisions. The result, the fe-male gametophyte (or embryo-sac), is a small structure containing eight haploid nuclei. Only one gametophyte is produced by each diploid megasporocyte; but, in a heterozygous plant, two mega-spores receive one allele, while the remaining two receive the other allele. It is a matter of chance which allele is present in the single survivor. In the $+/wx$ heterozygote this can be directly confirmed, since the iodine test shows that half the female gametophytes stain blue, the other half red.

Fertilization.—Three of the haploid nuclei present in the fe-male gametophyte take part in the further development. One of them is the egg-nucleus, and it fuses with one of the two sperm nuclei from the pollen, to form the initial diploid cell from which the new sporophyte of the next generation develops. The other sperm nucleus of the pollen also enters the female gametophyte at fertilization, and there fuses with *two* of the haploid nuclei de-veloped from the megaspore. The resulting triploid nucleus forms the beginning of the *endosperm,* which makes up a large part of the kernel of maize, but is not included in the mature sporophyte. The endosperm thus contains two maternal sets of chromosomes, one paternal. In the case of waxy, the iodine test shows that $+/+/+$, $+/+/wx$, and $+/wx/wx$ all stain blue, $wx/wx/wx$ stains red. That is, wx is recessive to $+$, even when one $+$ and two wx genes are present. It results from these relations that, if a $+/wx$ plant is pollinated by a wx/wx one, the kernels with normal and with waxy endosperm are equally numerous; if a $+/wx$ is self-pollinated the ratio is 3:1.

It will be seen that there is a correlation between the constitution of the endosperm and the embryo contained in the same seed. The two sperm-nuclei of a pollen grain are identical in composition, and the egg nucleus is identical with each of the maternal components of the endosperm. It must be expected, therefore, that kernels with normal starch will give rise to plants like the pure non-waxy strain $(+/+)$ or like the F_1 plants $(+/wx)$, those with waxy endosperm will give rise to plants like the pure waxy strain (wx/wx); this expectation is, in fact, realized in practice.

In this case, then, it is possible to use the iodine test to determine the constitution of each of the characteristic tissues, and to show that there is a detailed and exact step-by-step agreement between the cytologically observed behavior of the chromosomes and the distribution of the $+$ and wx genes.

Constancy of the Genes.—It is clear, in the examples given here and in Chapter II, that the two different alleles present in a heterozygote are recovered unchanged in the following generation. A recessive gene suffers no modification when it is carried for many cell-generations in an individual that shows the dominant phenotype. Another proof that the nature of the genes is not affected by the nature of the body that carries them has been furnished by certain experiments of Castle and Phillips. There is a white (albino) strain of guinea-pigs, that differs from fully colored strains by a single recessive gene, a. The ovaries of an albino female were removed, and those of a colored female were implanted in their place. The albino female was then mated to an albino male, and produced six young—all of them colored.

The implanted ovary was nourished by the blood of the albino female for from four to ten months, but the genes contained in it were not affected.

Evidence of this kind is sufficient to show that an influence of the body cells on the genes may be disregarded as a possibility in genetic experiments. Expressed in other words, which represent the traditional formulation, acquired characters are not inherited.

REFERENCES

Brink, R. A. 1929. Studies on the physiology of a gene. Quart. Review Biol. 4:521–543.

Matsuura, H. 1933. A bibliographical monograph on plant genetics. Hokkaido University, Sapporo.

Mendel, G. 1866. Versuche über Pflanzenhybriden. Verhandl. Naturforsch. Verein Brünn, 4.

Reprinted in Flora 89:364–403 (1901), and in Ostwald's Klassiker d. exakt. Wissensch. 121 (1901).

Translated ("Experiments on plant hybridization"):

1901. Journ. Royal Hort. Soc., 26.

1902. In Bateson, W., Mendel's principles of heredity, a defence. 212 pp. University Press, Cambridge.

1909, 1913. As appendix in Bateson, W., Mendel's principles of heredity. University Press, Cambridge.

1916, 1920. As appendix in Castle, W. E., Genetics and eugenics. Harvard University Press, Cambridge.

Also reprinted as separate pamphlet. 1938. 41· pp. Harvard University Press.

PROBLEMS

1. If the F_1 plants from green by yellow peas are crossed to the yellow parent, what colored seeds are produced? How would you determine the constitution of the resulting plants, and what types would you expect?

2. Maize is normally wind-pollinated. If seeds of pure strains of waxy and normal maize are planted in alternate rows, what kinds of seeds and what kinds of pollen grains are produced on the two types of plants?

3. Sugary endosperm (sweet corn) in maize is inherited as a simple recessive, *i.e.*, like waxy so far as the endosperm is concerned. There is, however, no visible effect in the pollen grains. Why is

it not desirable to plant sweet corn in plots adjacent to field corn (normal)?

4. If a breeder of Blue Andalusians wished to guarantee that only blue chicks would hatch from the eggs he sold, outline a breeding procedure he might follow. The proposed plan should cover three generations.

5. Rose comb in the fowl is dominant to single comb. The Wyandotte breeds are rose-combed, but many strains have the recessive allele for single comb present in them, with the result that occasional birds are homozygous for it. How would one eliminate the allele for single comb from a flock?

6. In human pedigrees normal individuals are shown by open figures, and abnormal ones by solid figures. In the following hypothet-

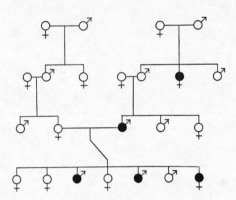

Hypothetical pedigree.

ical pedigree what is the probable inheritance of the character shown? So far as can be deduced, give the genotypes of the individuals shown in the pedigree.

7. Albinism in man is, at least in some instances, inherited as a simple recessive character. If a couple, each member of which is heterozygous for albinism, have four children, what is the probability that there will be three albinos and one normal?

8. As stated in the text, Mendel obtained 5,474 round to 1,850 wrinkled peas in the F_2 generation. With this number of individuals, calculate the numbers required for a perfect fit to a 3:1 ratio. What is the deviation from this? Calculate the standard error for this ratio with this number of individuals. Is this a satisfactory agreement between theory and observation? Explain.

9. In the F_2 generation of a cross between normal and tassel-seed (seeds in the staminate inflorescence) maize plants, 249 normal to 72 tassel-seed plants were obtained. What is the approximate probability that the deviation from a 3:1 ratio could be the result of errors of sampling?

CHAPTER IV

INDEPENDENT ASSORTMENT

MENDEL investigated the behavior of pea plants in which two pairs of genes were segregating. As previously described, the seeds may be either yellow or green (the corresponding genes being designated g^+ and g), and their shape may be round or wrinkled (the genes concerned here being designated w^+ and w). Mendel crossed a round yellow race ($w^+/w^+ \ g^+/g^+$ —usually written simply $w^+ \ g^+$) to a wrinkled green one ($w/w \ g/g$—or simply $w \ g$).

Two-character Segregation.—The F_1 seeds were, as expected, round and yellow, owing to the dominance of g^+ and w^+. In what follows it will be convenient to designate these as $w^+ \ g^+$, though in fact their formula is really $w^+/w \ g^+/g$. Their appearance is round yellow, though they are heterozygous. It is convenient to distinguish, in such cases, between the appearance of individuals, called their *phenotype,* and their genetic constitution, or *genotype.* In the present case, the F_1 seeds have the same phenotype as one of the parental strains, but differ from them in their genotype. It is often convenient to use the symbols for the genes to represent the phenotype. This has to be done with care, to avoid confusion; but it saves enough space (and often elaborate circumlocutions) to make it worth while. The convention is, in such cases, to write a single symbol for the dominant allele, if either one or two are present, and a single symbol for the recessive allele if the individual is homozygous for it. The cross outlined above would then be written, in terms of phenotypes:

$$w^+ \ g^+ \times w \ g \to w^+ \ g^+$$

Testcross Method.—Mendel studied the segregation in the doubly heterozygous F_1 plants by crossing them to plants homozygous for both recessive alleles—*i.e.,* to the wrinkled green strain. This method, which avoids the complications that arise when dominants are contributed by both parents, is known as the *testcross* method (sometimes the less exact term "*backcross*" is used, though it is properly employed only for a cross to one of the parental strains). In the present case, Mendel obtained the following results: when the F_1 was pollinated by *w g*—31 w^+ g^+, 26 w^+ *g*, 27 *w* g^+, 26 *w g*; when the *w g* plants were pollinated by the F_1—24 w^+ g^+, 25 w^+ *g*, 22 *w* g^+, 26 *w g*. It is clear that, in each case, the four possible classes of gametes were produced in equal numbers. Each of the two pairs of alleles segregated in a 1:1 ratio, and the two pairs were wholly independent, giving a 1:1:1:1 ratio when considered together (Fig. 18).

$g\,g\;w\,w$ ♀ X $g^+g\;w^+w$ ♂

EGGS	SPERMS: $g^+\;w^+$	$g^+\;w$	$g\;w^+$	$g\;w$
$g\;w$	$g^+g\;w^+w$ YELLOW ROUND	$g^+g\;w\,w$ YELLOW WRINKLED	$g\,g\;w^+w$ GREEN ROUND	$g\,g\;w\,w$ GREEN WRINKLED

Fig. 18.—Testcross of a double heterozygote in peas. The result is the same if the reciprocal cross (heterozygous female pollinated by recessive) is made.

Independent Assortment.—This principle of independent assortment (sometimes known as "Mendel's second law," segregation being considered the "first law") is of widespread application; it is, however, not universal—much of this book will be taken up with discussions of exceptions to it. Independent assortment is, in fact, the rule that applies to genes carried by separate chromosome pairs. There is cytological evidence, now to be considered, that the segregation of separate pairs at meiosis does itself follow the rule of independence.

Cytological Basis of Independent Assortment.—Carothers studied the meiotic divisions of a grasshopper, Brachystola magna, in which there was a heteromorphic pair of autosomes. One member of this pair was distinctly larger than its mate; the X, as usual in grasshopper males, had no mate. Carothers was able to

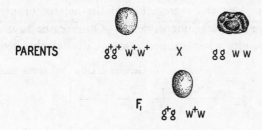

PARENTS g^+g^+ w^+w^+ X gg ww

F_1 g^+g w^+w

F_2 / EGGS	**SPERM** g^+w^+	g^+w	g w^+	g w
g^+w^+	g^+g^+ w^+w^+	g^+g^+ w^+w	g^+g w^+w^+	g^+g w^+w
g^+w	g^+g^+ w^+w	g^+g^+ ww	g^+g w^+w	g^+g ww
g w^+	g^+g w^+w^+	g^+g w^+w	gg w^+w^+	gg w^+w
g w	g^+g w^+w	g^+g ww	gg w^+w	gg ww

Fig. 19.—The results of crossing a round yellow-seeded pea by one with wrinkled green seeds.

determine that, when segregation occurred, the larger member of the heteromorphic pair passed to the same pole as the X in 154 of the cells examined, while in 146 cells the larger member of the pair and the X went to opposite poles. Here, then, is direct cytological proof that the segregation of the heteromorphic pair and the unpaired X was independent.[1] Other cases have since given similar results when studied cytologically.

			Genotypic ratio		Phenotypic ratio
		$\frac{1}{4}$ $\frac{g+}{g+}$ ——	$\frac{1}{16}$	$\frac{w+}{w+}$ $\frac{g+}{g+}$	
$\frac{1}{4}$	$\frac{w+}{w+}$	$\frac{1}{2}$ $\frac{g+}{g}$ ——	$\frac{2}{16}$	$\frac{w+}{w+}$ $\frac{g+}{g}$	
		$\frac{1}{4}$ $\frac{g}{g}$ ——	$\frac{1}{16}$	$\frac{w+}{w+}$ $\frac{g}{g}$	$\frac{9}{16}$ $w+$ $g+$
		$\frac{1}{4}$ $\frac{g+}{g+}$ ——	$\frac{2}{16}$	$\frac{w+}{w}$ $\frac{g+}{g+}$	
$\frac{1}{2}$	$\frac{w+}{w}$	$\frac{1}{2}$ $\frac{g+}{g}$ ——	$\frac{4}{16}$	$\frac{w+}{w}$ $\frac{g+}{g}$	$\frac{3}{16}$ $w+$ g
		$\frac{1}{4}$ $\frac{g}{g}$ ——	$\frac{2}{16}$	$\frac{w+}{w}$ $\frac{g}{g}$	
		$\frac{1}{4}$ $\frac{g+}{g+}$ ——	$\frac{1}{16}$	$\frac{w}{w}$ $\frac{g+}{g+}$	
$\frac{1}{4}$	$\frac{w}{w}$	$\frac{1}{2}$ $\frac{g+}{g}$ ——	$\frac{2}{16}$	$\frac{w}{w}$ $\frac{g+}{g}$	$\frac{3}{16}$ w $g+$
		$\frac{1}{4}$ $\frac{g}{g}$ ——	$\frac{1}{16}$	$\frac{w}{w}$ $\frac{g}{g}$	$\frac{1}{16}$ w g

Results of combining ratios for two segregating gene pairs.

The 9:3:3:1 Ratio.—From the results of Mendel's testcrosses it is possible to predict the results of selfing an F_1 round yellow pea plant. As figure 19 shows, random combinations of the four

[1] These observations were made at the first meiotic division. The relations here will be more easily understood after the discussion of meiosis in Chapter V.

kinds of eggs and four kinds of sperm give the nine possible genotypically different classes, in frequencies (per 16) ranging from 1 to 4. Collecting phenotypically like classes the result is: 9 $w^+ g^+$, 3 $w^+ g$, 3 $w g^+$, 1 $w g$. The numbers actually obtained by Mendel were: 315 $w^+ g^+$ (round yellow), 108 $w^+ g$ (round green), 101 $w g^+$ (wrinkled yellow), 32 $w g$ (wrinkled green) —obviously a close agreement with the expected 9:3:3:1.

$$\frac{3}{4} a^+ \begin{cases} \frac{3}{4} b^+ \begin{cases} \frac{3}{4} c^+ \quad — \quad \frac{27}{64} a^+b^+c^+ \\ \frac{1}{4} c \quad — \quad \frac{9}{64} a^+b^+c \end{cases} \\ \frac{1}{4} b \begin{cases} \frac{3}{4} c^+ \quad — \quad \frac{9}{64} a^+b\ c^+ \\ \frac{1}{4} c \quad — \quad \frac{3}{64} a^+b\ c \end{cases} \end{cases}$$

$$\frac{1}{4} a \begin{cases} \frac{3}{4} b^+ \begin{cases} \frac{3}{4} c^+ \quad — \quad \frac{9}{64} a\ b^+c^+ \\ \frac{1}{4} c \quad — \quad \frac{3}{64} a\ b^+c \end{cases} \\ \frac{1}{4} b \begin{cases} \frac{3}{4} c^+ \quad — \quad \frac{3}{64} a\ b\ c^+ \\ \frac{1}{4} c \quad — \quad \frac{1}{64} a\ b\ c \end{cases} \end{cases}$$

Results of combining phenotypic ratios for three gene pairs.

This ratio, 9:3:3:1, is characteristic for the F_2 when two independently inherited pairs of alleles are concerned. It is a phenotypic ratio when dominance is complete. If we combine algebraically the genotypic ratios for two independently segregating gene pairs, we have the results shown on page 54. As this shows, the genotypic ratio (which would be realized phenotypically also if both pairs of alleles gave distinguishable heterozygotes) is 1:2:1:2:4:2:1:2:1.

Either of these two methods may be used for working out the expectations from such crosses as these. The second, or algebraic, method is more convenient when one is dealing with more than two independent pairs of genes. Its use in a hypothetical case for three pairs, with complete dominance, is shown on page 55 (F_2 phenotypes only are indicated).

In general, if an individual is segregating for n independent genes, there will be produced 2^n kinds of gametes in equal numbers. On selfing, or mating together two such individuals, the completely homozygous recessive type (or any other complete homozygote) will constitute $(\frac{1}{4})^n$ of the resulting progeny.

MODIFIED RATIOS DUE TO INDISTINGUISHABLE CLASSES

The 9:3:4 Ratio.—The phenotypic ratios are subject to several types of modifications, due to resemblances between classes. One common type may be illustrated by a cross in guinea-pigs. The wild-type here has a reddish gray coat, called agouti. There exist several recessive mutant variations of this; one has a wholly black coat, another is an albino, with a white coat and unpigmented iris in the eyes. Using the symbols b and a to designate the genes concerned, the wild-type is $b^+/b^+\ a^+/a^+$, the black race is $b/b\ a^+/a^+$, the albino is $b^+/b^+\ a/a$. If, now, we cross a black by an albino, the F_1 animals have the constitution $b^+/b\ a^+/a$; since both + alleles are dominant, these animals are phenotypically wild-type. If they are mated together, the F_2 consists of 9 wild-type, 3 black, 4 albino (Fig. 20). This ratio arises from the fact that b^+/b^+, b^+/b, and b/b are indistinguishable in a/a animals—evidently because in such animals no pigment develops in the coat, and therefore the action of the b alleles cannot be detected. If the F_2 albinos are tested, by mating them to the pure black strain, it can be shown that this interpretation is correct; genotypically they fall into the classes b^+/b^+, b^+/b, b/b, in the expected ratio 1:2:1. The 9:3:4 ratio is, then, a 9:3:3:1 ratio in which the last two classes are phenotypically alike.

The 9:7 and 27:37 Ratios.—In sweet peas there are two kinds of recessive white flowers, each breeding true on selfing; the two

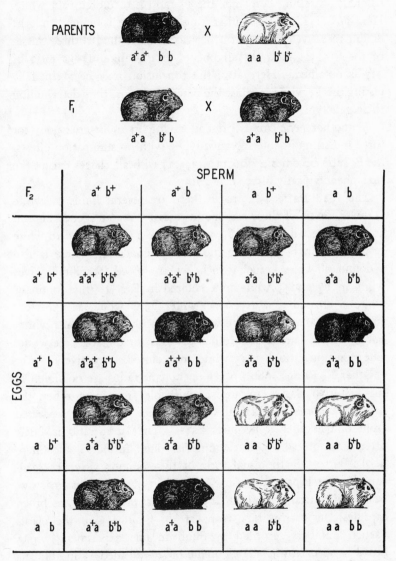

Fig. 20.—The results of crossing a black guinea-pig by an albino of a particular strain.

kinds are indistinguishable. When they are crossed, the F_1 is colored; on selfing F_1 plants, the F_2 ratio is 9 colored to 7 white. Here it is evident that either single recessive is white, and the double recessive cannot go further; accordingly the last three classes of the $9:3:3:1$ ratio are all phenotypically alike, and a ratio of $9:7$ is obtained. Here also, the conclusion may be verified by testing the F_2 whites against the parental strains, thus determining their genotypes.

If another gene-pair is present, giving an F_1 heterozygous for three indistinguishable recessive types, with no summation effects, the F_2 ratio becomes $27:9:9:3:9:3:3:1$, with all classes except the first phenotypically alike, i.e., $27:37$.

The 13:3 Ratio.—In maize there are several kinds of white kernels. Some of these are due to recessive mutant genes, and on crossing give wild-type in F_1, with $9:7$ or $27:37$ colored to white respectively in F_2 just as in the cases enumerated. There is also a dominant white kernel, which, when crossed to colored, gives white in F_1 and 3 white to 1 colored in F_2. If this dominant white is crossed to a recessive white, F_1 is white, as expected. In F_2 there are 3 dominant white to 1 without dominant white; but ¼ of this latter class are homozygous for the recessive white. The resulting ratio is $13:3$ white and colored, respectively.

Other F_2 Ratios.—Still other phenotypic ratios are possible. If the first two classes are indistinguishable from each other, the $9:3:3:1$ becomes $12:3:1$; if the first three cannot be separated from one another it becomes $15:1$. The latter ratio is fairly often encountered. One example is found in Bursa, the Shepherd's Purse. In the usual form of this common weed the seed-capsules are broad and flat. There is a race that has narrow, nearly cylindrical, capsules. If the two are crossed, F_1 has broad capsules; the narrow type reappears in F_2 in $\frac{1}{16}$ of the plants (actual numbers recorded by Shull in one experiment, 7,943 broad : 506 narrow). Tests of the broad-capsuled F_2 individuals show that the interpretation is correct; many of them, being homozygous for one of the two recessives and heterozygous for the other, give $3:1$ ratios in F_3.

This type of case, where a phenotypic effect is produced only by the simultaneous homozygosis of two recessives, sometimes occurs in the form where still another recessive must be homozygous. In this case the F_2 ratio for a triple heterozygote, $27:9:9:3:9:3:3:1$, is modified by the fact that the first seven classes as listed above cannot be distinguished from one another; the resulting ratio is $63:1$.

Distinguishing Different Ratios.—Some of the ratios described above are not markedly different, though they result from different types of phenotypical complications. In practical work it is often difficult to distinguish, for example, $9:7$ from $27:37$, $13:3$ from $3:1$, or $15:1$ from $63:1$. In such cases the rearing of large families is desirable, for this will reduce the error due to random sampling—increase the probable significance of a deviation from one of the possible ratios. In general, however, where it is possible to do so, it is better to test the F_2 individuals or to make a testcross of the F_1, since the information from such crosses is likely to give a more direct and critical test of the genotypical constitution of the population. Limitations of time, space, or expense, however, often make it necessary to be as sparing as possible of such tests.

OTHER KINDS OF MODIFIED RATIOS

There exist other reasons for preferring tests of individuals rather than increasing the size of the F_2 families in cases where the significance of the ratios is doubtful. We have so far taken into account only the effects of random sampling as leading to distortion of ratios; but at least two other possible causes of distortion must usually be reckoned with—misclassification of individuals, and differential viability.

Misclassification.—It is often desirable to work with types that can be distinguished only with difficulty, in which case some individuals may be classified wrongly. In other words, if the range of variation of the phenotypes overlaps, the ratios obtained may be misleading. In such cases tests of the genotypical constitution of individuals—breeding tests—must be resorted to.

Differential Viability.—Some types are less vigorous than others, with the result that, even if a given ratio was exactly realized at the time of fertilization, the death of a greater proportion of one class than of another may seriously distort the relative frequencies before one records the results. There are three ways in which this difficulty can be minimized: in some cases it is possible to estimate the total mortality (by counting eggs or seeds, and comparing with the number of individuals present when classifications are made), thus setting a limit on the possible distortion of the ratios; sometimes it is possible to decrease the total mortality by improving culture conditions, thereby presumably also decreasing the *differential* mortality; the third method is to make tests of the surviving individuals. The latter method may be illustrated by a hypothetical example.

Suppose one crosses two races, and in F_2 obtains a ratio near 15:1; suppose further that the recessive class is suspected of having low viability. The problem is, does this represent a true 15:1 ratio, or is it a 3:1 distorted by low viability of the recessive class? In other words, are we concerned with one pair of genes, or with two? In general, the simplest test here is to determine the constitutions (by selfing if possible) of the F_2 plants belonging to the dominant class. If we are dealing with a single pair of genes, one third of them will breed true and the other two thirds will segregate recessives in the same proportion as did the F_1 when selfed. If the true ratio is 15:1, the F_2 dominants will fall into three classes: $\frac{7}{15}$ will breed true, $\frac{4}{15}$ will give a 15:1 ratio, $\frac{4}{15}$ will give 3:1. These relations can readily be worked out by means of the checkerboard method. Such a test, giving information on the *genotypical* constitution, will always be more satisfactory than will the same number of individuals classified only by *phenotypes*.

REFERENCES

Babcock, E. B. and R. E. Clausen. 1927. Genetics in relation to agriculture. 673 pp. McGraw-Hill Co., New York.

Matsuura, H. 1933. A bibliographical monograph on plant genetics. Hokkaido University, Sapporo.

PROBLEMS

1. If an F_2 is segregating for a single gene, what is the probability of a given individual being homozygous recessive? If two genes are segregating, what is the probability of an F_2 individual being homozygous recessive for both of them? In the same way, if three and four genes are segregating, what is the probability of an F_2 individual being homozygous recessive for *three* and *four* genes respectively? What is the general rule for the probability of obtaining an F_2 individual homozygous recessive for *n* genes?

2. Repeat the above problem substituting heterozygous for homozygous recessive.

3. In an F_2 segregating for genes *a* and *b*, what is the probability of obtaining each of the following genotypes:

    a. $+ + + +$
    b. $+ + + b$
    c. $+ + b b$
    d. $+ a + b$
    e. $a a + b$

4. In an F_2 segregating for three genes *a*, *b*, and *c*, what are the probabilities of obtaining individuals of the following phenotypes:

    a. $+ + +$
    b. $+ + c$
    c. $a b +$

    What are the probabilities of obtaining the following genotypes:

    a. $+ + + + + +$
    b. $+ + + b + c$
    c. $+ a + b + c$
    d. $+ a + b + +$

5. In a self-fertilizing plant, what proportion of the F_2 individuals from an F_1 heterozygous for three genes will breed true in F_3?

6. It is possible to obtain the same F_1 and F_2 results, with guinea-pigs, as are described in the text and illustrated in figure 20, using as original parents an agouti and an albino. What genotypes should such agoutis and albinos have?

7. What phenotypes and in what relative frequencies would be expected in F_1 and F_2 from a cross of a white-eyed Drosophila female by a stubbloid male? (White is a sex-linked recessive

and stubbloid is an autosomal recessive, and each original strain carries the dominant wild-type allele of the mutant present in the other.)

8. Black body and scarlet eyes are independent autosomal characters in Drosophila. An F_1 female from the cross black $\times$ scarlet is mated to a male from the cross black $\times$ black scarlet. What phenotypes are expected among the offspring and in what relative frequencies are they expected?

9. There are three true-breeding strains of maize: one with white aleurone, one with purple, and one with red aleurone. White $\times$ purple gives white in F_1 and three white to one purple in F_2; red $\times$ purple gives purple in F_1 and three purple to one red in F_2. What would you predict for F_1 and F_2 on crossing white and red?

10. In F_2 of a cross of a white-flowered plant by a purple-flowered plant, there are 380 purple to 100 white-flowered offspring. What is the probability of getting a deviation as great or greater than this from a 3:1 ratio as a result of sampling errors? Of the other ratios given in the text, which ones could fit these data? Assuming that the plant concerned can be self-pollinated, and that, in the following year, space is available for growing only 500 plants, how would you differentiate between the various possible interpretations of the F_2 results?

11. Two true-breeding types of maize with colorless aleurone give colored aleurone in F_1 and 9 colored to 7 colorless kernels in F_2. Remembering that a given plant can be used as a pollen parent in several crosses and can also, if desired, be self-pollinated as well as outcrossed, how would you distinguish the different genotypes in the F_2 generation? Remember that the parental strains can be made use of in tests.

CHAPTER V

LINKAGE

IN previous chapters we have considered simple cases of Mendelian inheritance; those in which no more than a single gene pair for each chromosome pair was followed. It was early realized that since organisms have a limited number of chromosomes, the number of independently segregating gene pairs must be limited accordingly. With the active study of heredity following the rediscovery of Mendel's paper, it was soon found that the number of characters that individually followed Mendelian principles was greater than the number of chromosome pairs to which the differentiating genes could be assigned. For a number of years this apparent discrepancy was advanced as an argument against the chromosome theory of heredity. The proponents of the theory, however, early pointed out that this argument was valid only so long as the principle of independent assortment held. With the rapidly increasing body of data concerning inheritance of simple characters, it became evident that independent segregation was in fact far from universal in its application.

Purple-vestigial Crosses.—The eye color mutant known as purple (pr) in Drosophila is a simple autosomal recessive character. A small wing character, called vestigial (vg—Fig. 21), is likewise a simple autosomal recessive. Unlike previously considered cases of more than one character, purple and vestigial are not inherited independently. By means of special techniques which will be considered later, it is nevertheless possible to have both of these characters represented in a single individual. If a female that is both purple and vestigial ($pr\ vg$) is crossed to a wild-type male, the sons and daughters are all wild-type. This is the result we would expect, since crosses in which these two characters are involved separately show that the wild-type parent must contribute a dominant allele of the purple gene

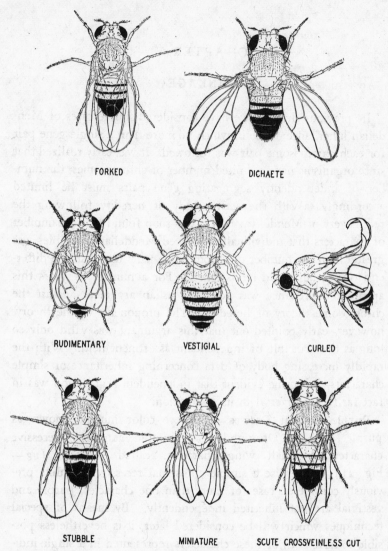

FORKED   DICHAETE

RUDIMENTARY   VESTIGIAL   CURLED

STUBBLE   MINIATURE   SCUTE CROSSVEINLESS CUT

Fig. 21.—Various wing and bristle characters in Drosophila. (Drawn by
E. M. Wallace.)

and also a dominant allele of the vestigial gene. If the F_1 males,
from the cross purple vestigial female by wild-type male, are
tested by crossing to double recessive females genetically like the

purple vestigial parent (*i.e.*, the testcross method), there result, instead of the four types expected with independent segregation, two types only, in equal numbers, and these are like the original parents, *i.e.*, purple vestigial and wild-type.

An actual experiment of this type made by Bridges gave the following result:

$$\text{Parents } pr^+ \ vg^+ \ ♀ \ \times \ pr \ vg \ ♂$$

$$F_1 \quad \frac{pr^+ \quad vg^+}{pr \quad vg} \quad ♀♀ \text{ and } ♂♂$$

Testcross—*pr vg* ♀ × F_1 ♂
wild-type ($pr^+ \ vg^+$) 519
purple vestigial (*pr vg*) 552
vestigial ($pr^+ \ vg$) 0
purple ($pr \ vg^+$) 0

(Note: A double male or female sign, *i.e.*, ♀♀, is a plural form.)

It will be noted that in this example the original cross is the reciprocal of that previously indicated. In this original cross it makes no difference in the final result whether the double recessive is used as the male or the female parent.

In this cross, it is evident that the *pr* gene remains with the *vg* gene during the process of meiosis. In the same way, the two wild-type alleles, pr^+ and vg^+, remain together as they were in the original parent.

Parental Combinations Recovered.—If the doubly heterozygous F_1 male is obtained from a cross *vg* ♀ × *pr* ♂ (the reciprocal, *pr* ♀ × *vg* ♂ would give the same result) and is similarly tested by crossing with a *pr vg* ♀, the result is different. In an actual test (Bridges) such an experiment gave results as follows:

$$P \quad pr^+ \ vg \ ♀ \ \times \ pr \ vg^+ \ ♂$$

$$F_1 \quad \frac{pr^+ \quad vg}{pr \quad vg^+} \quad ♀♀ \text{ and } ♂♂$$

Testcross—*pr vg* ♀ × F_1 ♂
wild-type ($pr^+ \ vg^+$) 0
purple vestigial (*pr vg*) 0
vestigial (*vg*) 358
purple (*pr*) 346

Here, again, there result two classes only, and these in equal numbers, but now the two classes are purple and vestigial. The first experiment shows that where the genes *pr* and *vg* are originally together they remain together, and the second shows that where they were originally separate, the parental combinations remain together just as persistently. This phenomenon is known as *linkage*. Several explanations of linkage have been advanced in the past, but it is now quite clear that the correct interpretation is that genes are linked because they are carried in the same chromosome pair. This may be represented schematically in the following way:

$$P \quad \frac{pr \quad vg}{pr \quad vg} \quad ♀ \quad \times \quad \frac{pr^+ \quad vg^+}{pr^+ \quad vg^+} \quad ♂$$

$$F_1 \quad \frac{pr^+ \quad vg^+}{pr \quad vg} \quad ♀♀ \text{ and } ♂♂$$

$$\text{Sperm formed by } F_1 \; ♂ \; — \; \underline{pr^+ \quad vg^+} \text{ and } \underline{pr \quad vg}$$

Here, following a convention previously indicated, chromosomes are represented by horizontal lines. Here, too, the formula is usually simplified by omitting the second horizontal line, and, further, by dropping the bases that distinguish one wild-type allele from another. This does not introduce any confusion since each $+$ is placed immediately above or below the symbol for the corresponding mutant gene. Following these simplifications we represent the above F_1 by the formula $\frac{+\quad+}{pr \quad vg}$. The other possible arrangement, namely that in which the recessive alleles are in different homologs, is indicated by the formula $\frac{+\quad vg}{pr \quad +}$.

It is sometimes convenient to specify briefly one or the other of these two arrangements of genes and for this purpose the terms *coupling* and *repulsion* are often used. Coupling signifies that the two recessive alleles are carried in one chromosome and the two dominant alleles in the other. This is so without regard to whether one or both mutant alleles are dominant. These terms are based on a now universally abandoned interpretation of linkage and therefore have historical significance only. They are, however, met with rather frequently in genetical literature.

INCOMPLETE LINKAGE

The examples given above illustrate a special case known as complete linkage. The essential characteristic of this simple case is that only parental combinations of genes are represented among the gametes of the F_1 heterozygote. In other words, chromosomes are transmitted from parent to offspring as units. Complete linkage is by no means the general rule. As a matter of fact, if we repeat the experiments described above, using F_1 females rather than males in the testcross, the results will be quite different. Testing F_1 daughters from the mating wild-type female by purple vestigial male, Bridges obtained the following results:

$$P \qquad \frac{+\quad+}{+\quad+} \; \female \; \times \; \frac{pr \quad vg}{pr \quad vg} \; \male$$

$$F_1 \qquad \frac{+\quad+}{pr \quad vg} \; \female\female \; \text{and} \; \male\male$$

$$\text{Testcross} \quad \frac{+\quad+}{pr \quad vg} \; \female \; \times \; \frac{pr \quad vg}{pr \quad vg} \; \male$$

+	+	1,339
pr	vg	1,195
+	vg	151
pr	+	154
	Total	2,839

Here the four possible phenotypes for these two characters are all represented. There are, however, about nine times as many flies phenotypically like the parents as flies of *new combinations*. Here then there is a tendency for the genes to remain in the combinations in which they were associated in the parents; but this tendency, unlike that in the case of gamete formation by a male of identical genetic constitution for these gene pairs, is not complete. The proportion of new combinations is usually expressed as a percentage of the total. In the example just given it is $305/2839 \times 100$, or 10.7 per cent.

Parental Combinations Most Numerous.—If a testcross is made using the F_1 female of the constitution $\dfrac{+\quad vg}{pr\quad +}$ and a homozygous *pr vg* male, results similar to the following (from Bridges) will be obtained:

$$\text{P} \qquad \dfrac{+\quad vg}{+\quad vg}\ \text{♀} \times \dfrac{pr\quad +}{pr\quad +}\ \text{♂}$$

$$\text{F}_1 \qquad \dfrac{+\quad vg}{pr\quad +}\ \text{♀♀ and ♂♂}$$

$$\text{Testcross:}\ \ \text{F}_1\ \text{♀} \times \dfrac{pr\quad vg}{pr\quad vg}\ \text{♂}$$

+	+	157
pr	vg	146
+	vg	965
pr	+	1,067
	Total	2,335

Here, again, the parental combinations are more numerous than are the new combinations. In this example 13.0 per cent of the individuals from the testcross are new combinations. This percentage is not very different from that obtained when the *pr* and *vg* genes were originally in the same chromosome.

Many experiments similar to those just described have been made and it is a general rule that for the two gene pairs *pr* and *vg*, regardless of the arrangement in F_1, the percentage of new combinations is constant within reasonable limits. In practice this relative uniformity is attainable only by keeping genetic and environmental factors reasonably constant.

Linkage between Sex-linked Genes.—It has been indicated above that linkage results when two pairs of alleles are carried in a single pair of chromosomes. We have already seen (Chapter II) reasons for supposing that sex-linked genes are carried in the X chromosomes; if these conclusions are correct, it should follow

that different sex-linked genes are linked to each other. The case
now to be considered shows that this deduction is correct.

If a female homozygous for the two sex-linked recessive genes
yellow body (y) and white eye (w) is crossed to a wild-type
male, the daughters will be phenotypically wild-type, and genotypically heterozygous both for y and for w; the sons will be $y\ w$.
This cross can be represented schematically as follows:

$$P \quad \frac{y \quad w}{y \quad w}\ ♀ \ \times \ +\ +\ ♂$$

$$F_1 \quad \frac{+\ +}{y \quad w}\ ♀♀ \text{ and } y\ w\ ♂♂$$

It should be remembered that males have only one X chromosome plus a Y chromosome. The Y chromosome is frequently
represented in formulae as a hooked horizontal line ($\longrightarrow$),
but in most cases it is simpler to omit this. We have already
pointed out (Chapter II) that the Y chromosome does not carry
alleles of ordinary sex-linked genes.

Since, in the cross indicated, the F_1 males are all yellow white
($y\ w$), mating of F_1 females and males to give the F_2 generation
is the equivalent of a testcross, and both females and males of the
F_2 can be used as a measure of gametic frequencies. Actual
crosses of this type made by Dexter gave in F_2:

+	+	8,093
+	w	93
y	+	81
y	w	6,672

Total 14,939

New Combinations 174 = 1.2 per cent

Here the proportion of new combinations is lower than that for
the autosomal genes pr and vg.

The yellow female by white male cross gives wild-type daugh-
ters $\left(\dfrac{y\ +}{+\ w}\right)$ and yellow sons ($y\ +$). Here the F_2 results are
not the equivalent of those from a testcross since the F_2 females

each receive an X chromosome from the F_1 yellow males, and all will consequently be phenotypically not-white. Nevertheless it is not essential to make a special testcross to measure the frequency of recombinations. The F_2 males get their Y chromosomes from the F_1 males, and their phenotypes are a direct indication of the genotypes of the gametes produced by the F_1 female. To clarify

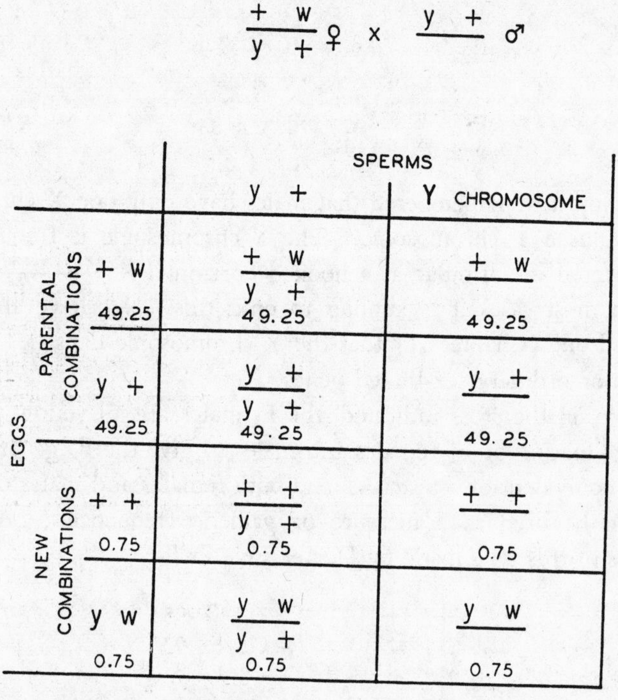

$$\frac{+\quad w}{y\quad +}\ ♀\ \times\ \frac{y\quad +}{}\ ♂$$

		SPERMS	
		y +	Y CHROMOSOME
PARENTAL COMBINATIONS	+ w 49.25	$\frac{+\quad w}{y\quad +}$ 49.25	+ w 49.25
	y + 49.25	$\frac{y\quad +}{y\quad +}$ 49.25	$\frac{y\quad +}{}$ 49.25
NEW COMBINATIONS	+ + 0.75	$\frac{+\quad +}{y\quad +}$ 0.75	+ + 0.75
	y w 0.75	$\frac{y\quad w}{y\quad +}$ 0.75	$\frac{y\quad w}{}$ 0.75

(EGGS)

Fig. 22.—Checkerboard diagram showing the genotypes and the frequencies with which they are expected in the F_2 generation of the cross $y + ♀ \times + w ♂$ in Drosophila.

these relations, we can make use of a checkerboard diagram (Fig. 22). It is obvious, however, that the special testcross of the F_1 females to $y\ w$ males is more efficient, since both males and females can then be used as a measure of relative gametic frequencies. From an experiment of this type Morgan and Cattell obtained the following results:

$$F_1 \qquad \dfrac{+ \quad w}{y \quad +}$$

Testcross F_1 ♀ $\times$ $y\ w$ ♂

+	+	86
+	w	4,292
y	+	4,605
y	w	44

Total   9,027

New combinations   130 = 1.4 per cent

Again parental combinations are greatly in excess of recombinations, there being only 1.4 per cent of the latter. This percentage is very close to that obtained in the first cross involving yellow and white.

Percentages of New Combinations.—The percentages of new combinations obtained for the experiments with y and w, 1.2 and 1.4, are considerably lower than the corresponding percentages for pr and vg. We could have selected other examples where the percentage approximated any value from 0 to 50 per cent. The significance of the relative constancy for a particular combination of two gene-pairs and of the differences between different combinations of gene-pairs will be considered in detail in the next chapter.

MEIOSIS

As we shall see, it turns out that new combinations of linked genes, known technically as *recombinations,* are of great interest to the student of genetics. In order to appreciate their significance and understand how they are produced it is necessary to understand the essential features of the process of meiosis, the two divisions of which are sometimes referred to as the "maturation divisions" (animals) or "reduction divisions."

Mitosis.—In order to emphasize the significant features of meiosis it will be useful to recall the characteristics of mitosis, or somatic cell division. As has been pointed out, chromosomes are present in pairs in the somatic cells of higher plants and animals, one member of each pair originally derived from one parent, the

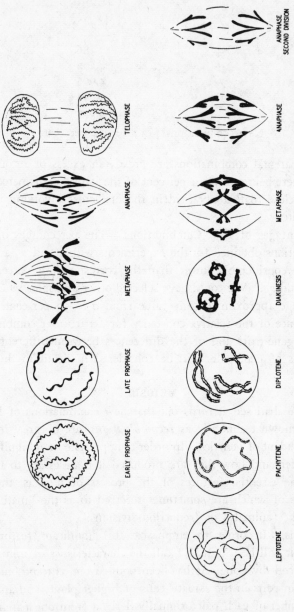

Fig. 23.—Semi-diagrammatic representations of mitosis (above) and meiosis (below). Telophase of the first meiotic division, interkinesis, and prophase of the second meiotic division are omitted. Telophase of the first meiotic division, interkinesis, and prophase of the second meiotic division are omitted. Internal structure of chromosomes is not indicated.

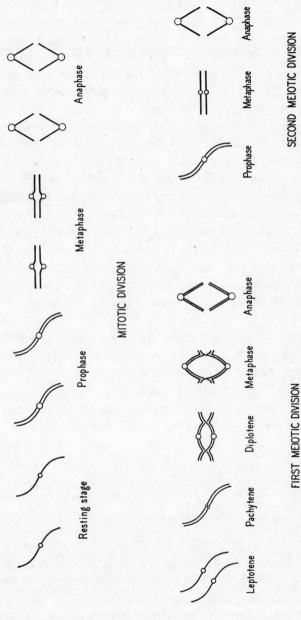

Fig. 24.—Schematic representation of the essential features of mitosis (above) and of meiosis (below). Coiling of chromatids is omitted. Centromeres are represented by open circles.

other member from the other parent. With each division of the cell into daughter cells, there is a longitudinal division of each chromosome. One daughter half goes to one daughter cell, the sister half to the sister cell. This briefly describes the precise mechanism by which chromosomes are transmitted from one cell generation to the next. This process is shown diagrammatically in figures 23 and 24.

Chromosome Conjugation.—In contrast to their behavior in mitosis, the two members of a pair of chromosomes come to lie side by side in the early stages of the first meiotic division. This pairing of homologous chromosomes in meiosis, a process known as *conjugation* (or synapsis), may be thought of as replacing the longitudinal division of mitotic chromosomes. It can be seen, by carefully studying the chromosomes in the prophase of the first meiotic division, that the chromosomes are long and thread-like before conjugation and that they are often longitudinally differentiated with various sized enlargements called *chromomeres* (Fig. 23). This stage is called *leptotene.* During the process of conjugation (stage known as *zygotene*), like parts of homologs come to lie side-by-side; apposed chromomeres are morphologically similar. The essentials of this process, then, are that like chromosomes pair, and that this pairing is regionally specific. The stage during which the thread-like chromosomes are associated in pairs is known as *pachytene.* During pachytene the chromosomes divide longitudinally to give four parallel threads, each known as

---

Plate II

A, B, C, and D. First meiotic division in the onion, Allium cepa. A, Pachytene. B, Diplotene. C, Diakinesis. D, Metaphase. (Photographs by courtesy of Professor Karl Sax.)

E, F, G, and H. Pachytene chromosomes of maize. E, Entire nucleus of plant with one extra chromosome (*i.e.,* 2n + 1). In a preparation such as is shown here, each chromosome pair can be identified by careful study. F, Photograph especially taken to illustrate details of chromosome 6, which is below and attached to the nucleolus. This attachment to the nucleolus at a specific place in the chromosome is a characteristic of this particular chromosome. G, Detail of chromosome 9. Note the dark staining knob at the end of the short arm. H, A conjugated pair of B-type chromosomes. Note characteristic structure and dark staining regions. (Photographs by Dr. Barbara McClintock.)

PLATE II

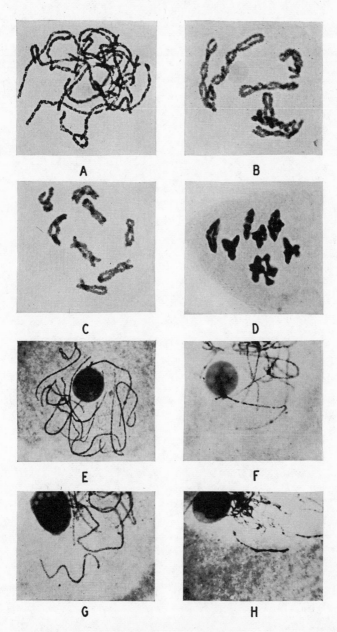

For description, see facing page.

a *chromatid*. There is, in each chromosome, a localized region that does not ordinarily take up the dyes with which chromosomes are usually stained. This region remains single for some time after the rest of the chromosome has divided. It is known as the *centromere* (or spindle attachment, insertion region, or kinetochore).

Chiasmata and Crossing Over.—The pachytene division results in the presence of four closely apposed chromatids; almost immediately there is a separation of chromatids by pairs, resulting (at any one region) in two groups, each containing two strands. Various kinds of indirect evidence indicate that *sister chromatids* remain together, *i.e.,* the separation is along the plane of conjugation (Fig. 25). The chromatids tend to move apart as though the two pairs were repelling each other. In the typical case, the moving apart is arrested by changes of pairing partners among the chromatids. There may be one or many such changes of partners; because of the resulting cross-like appearance these changes of partners are called *chiasmata* (singular, *chiasma*). In order to simplify the treatment, we shall consider in this chapter only the case in which one chiasma is formed between a pair of homologous chromosomes. Chiasmata result from an interchange of corresponding segments of homologous chromatids (Fig. 25). As will be demonstrated in detail later, only two of the four chromatids undergo such an interchange of parts, at any one point. The new chromatids formed as a result of this interchange are known genetically as crossover chromatids or more briefly as *crossovers*. For the time being crossovers may be considered to be the same as recombinations; a distinction between the two terms will be made in the next chapter. The process by which crossovers are produced is known as *crossing over*. Through indirect evidence, many of the characteristics of crossing over are known, but very little is known about *how* it occurs.

Diplotene.—The stage of meiosis immediately following the opening out of the four chromatids which make up the chromosome *tetrads,* is known as *diplotene*. The chromosomes have, at this stage, become shorter and thicker. This shortening is largely or wholly the result of coiling of the chromatids. This process of coiling and consequent shortening of the chromosomes con-

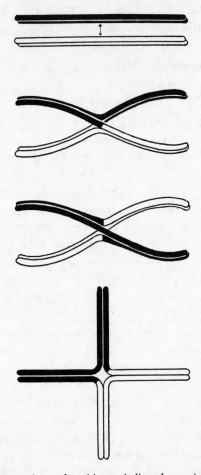

Fig. 25.—Different views of a chiasma indicated as arising as a result of crossing over between two of four chromatids. It is suggested that the use of models made of clay, wax, or other suitable material, will be found to help in visualizing chiasmata. It is important, in understanding subsequent two-dimensional figures, to appreciate that the three views shown here represent essentially the same thing.

tinues and diplotene merges imperceptibly into the next stage known as *diakinesis*.

Terminalization of Chiasmata.—During diplotene and diakinesis there is often a progressive reassociation of paired chromatids such that a chiasma "moves" away from the centromere toward the distal end of the chromosome pair (Fig. 26). This movement is called *terminalization*. Complete terminalization of chiasmata is characteristic of some species but not of others. Maize, for example, shows partial terminalization of chiasmata

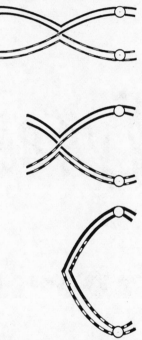

Fig. 26.—Diagrammatic representation of terminalization of a chiasma at successive stages of meiosis.

(Fig. 27). The evening primrose, Oenothera, on the other hand, is characterized by complete terminalization (Fig. 80). Following complete terminalization of chiasmata, homologs are held together by thread-like connections joining their homologous ends. The precise nature of these terminal connections is not understood.

Maximum contraction of the first meiotic division chromosomes is attained at metaphase. Here the chromosome *bivalents* (tet-

Fig. 27.—Successive stages of the first meiotic division in Zea mays. From the top: diplotene, early diakinesis, late diakinesis, and metaphase. This is an example of partial terminalization of chiasmata. The leftmost unit is a complex of four chromosomes that is of no consequence for purposes of this figure. The vertical axis in each unit of the three upper rows corresponds to the horizontal axis in the lower row. (From Darlington.)

rads in terms of chromatids) are oriented on the equatorial plate, following which the centromeres begin to move toward the poles of the spindle. The shape of a given bivalent at metaphase depends on the number and position of chiasmata (*i.e.*, the amount of terminalization) and on the length of the chromosomes.

Anaphase and the Second Division.—The movement of chromosomes toward the poles, anaphase, follows metaphase. Any chiasmata that have not already terminalized by the late metaphase are pulled apart at anaphase. This pulling apart simply reassociates the chromatids; there is no breakage of chromatids.

The daughter nuclei resulting from the first meiotic division go into a stage known as interkinesis, which is somewhat like a short resting stage in mitosis. The centromeres are still undivided at this stage, so that two chromatids have a common centromere. The second meiotic division usually follows the first closely. The chromosomes, now the haploid number, each made up of two chromatids, come to lie on the plate of the second division spindle. The centromeres divide and a single chromatid from each unit of two goes to each of the daughter cells.

Reduction in Number of Chromosomes.—The four nuclei resulting from the two meiotic divisions each have the haploid number of chromosomes, half the number characteristic of the meiotic mother cell. In other words, the net result of the two divisions is to reduce the chromosomes from two sets to one set.

Other Interpretations.—Before continuing with the genetic consequences of meiosis, it should be pointed out that the account of mitosis and of meiosis given above does not apply to all organisms; there are many modifications of the process. Furthermore, for the sake of brevity and simplicity, we have presented only one interpretation. There are a number of points concerning which cytologists have no unanimity of opinion. For example, it is very difficult to determine by direct observation just when chromosome threads divide longitudinally, and many investigators maintain that such division occurs at different stages of the life cycle from those indicated in our account. We have implied in the above account that homologous chromosomes are held together through diplotene, diakinesis, and metaphase by chiasmata. This is one interpretation of what is seen cytologically. It may not be entirely correct and it is certainly not universally applicable (*e.g.*, the chromosomes of the male of Drosophila very clearly remain paired during the first meiotic division without chiasmata). The indicated relation between physical interchange of corresponding segments of homologous chromosomes, genetic crossing over, and chiasmata known as the *chiasmatype theory*, is the best hypothesis available. It can be argued that complete proof of this relation has not been made. One of the difficulties of getting evidence concerning this relation is the fact that chromosomes are very small and must be studied at very high magnification. In many organisms chiasmata are difficult to identify with certainty; they may, for example, be confused with simple overlaps of homologous chromosomes. Any extensive discussion of such controversial questions as those just indicated is beyond the scope of this book; these subjects belong more properly to cytology and the reader is referred to the standard accounts cited at the end of the chapter for fuller treatments.

CROSSING OVER

Returning to the question of genetic crossing over or the production of new combinations of linked genes, we have seen that for a single chiasma, there are two crossover and two non-crossover chromatids (Fig. 25). Following the segregation of these four chromatids there will be two nuclei containing crossover chromatids and two containing non-crossovers. In general, we are able to recover for genetic study only a single product of meiosis. It is true, of course, that in the male of most animals all four products give rise to sperms and in higher plants all four microspores give rise to pollen grains. The difficulty is that in general there is no way of telling which four sperms or pollen grains came from the same meiotic division, and deductions must be made on a statistical basis. In female animals and in pistillate parts of higher plants it is of course the general rule that only a single product of meiosis has a functional future.

Chiasmata and Recombination.—If we now assume a chromosome pair to be heterozygous for two gene-pairs, say cinnabar (an eye-color) and vestigial, it is evident that in those meiotic divisions in which no chiasma is formed by this chromosome pair *between the loci of cinnabar and vestigial,* only parental combinations for these two genes will result, two of each of course. But in those meiotic divisions in which a chiasma is formed between cinnabar and vestigial there will result two new

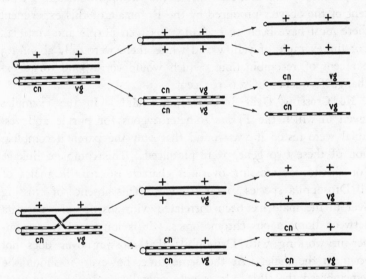

Fig. 28.—Schematic representations of meiotic divisions with (below) and without (above) a chiasma between the genes cinnabar and vestigial of Drosophila. For simplicity only one arm of the chromosome pair is shown. Centromeres indicated with circles.

combinations, all four of the possible types being present (Fig. 28). If chiasmata were formed between the loci of cinnabar and vestigial in 20 per cent of the oöcytes but not in the remaining 80 per cent, we can readily calculate the relative proportions of parental and new combination of genes that would result. Of one hundred oöcytes, the 80 with no chiasmata give 80 eggs, statistically 40 of one parental combination and 40 of the other; the 20 oöcytes, each with a chiasma between cinnabar and vestigial, will give 20 eggs, and, again statistically, half of these will be parental combinations, 5 of each. The other half will be new combinations or crossovers, 5 of each of the two possible types, or a total of 10 per 100 eggs (10 per cent). It is clear from this that the crossover frequency is equal to one half of the *exchange frequency* (sometimes known as *chiasma frequency*). In the example given earlier in this chapter, there were about 12 per cent of new combinations for the linked genes purple and vestigial. This means that in approximately 24 per cent of the oöcytes produced by the F_1 female of this experiment there must have been an exchange between purple and vestigial. For the example of yellow and white, there were only about 1.3 per cent of recombinations, which would correspond to an exchange frequency of 2.6 per cent.

No Crossing Over in Drosophila Male.—In the examples given in which the F_1 *males* heterozygous for purple and vestigial were tested it was found that only the parental combination of these two genes were produced. This complete linkage (or absence of crossing over) is characteristic of the males of all Drosophila species so far studied. This absence of crossing over in the male has been correlated with absence of chiasmata between homologous chromosomes. It is important to remember in working with Drosophila that crossing over does not occur in the male. At the same time, however, it should be remembered that this rule is not generally applicable. In fact, among the organisms ordinarily studied by geneticists, Drosophila is the only one in which crossing over is strictly limited to one sex.

Linkage and Crossing Over in Maize.—In maize, crossing over is known to occur in both megasporocytes and microsporocytes. Furthermore the frequencies are usually nearly the same in the two types of mother cells. As an example, the gene C (colored aleurone) is in the same chromosome pair as the gene *wx* (waxy endosperm). The cross $\dfrac{C\ +}{+\ wx}$ ♀ $\times$ $+\ wx$ ♂ gives results similar to the following (from Kempton):

C	$+$	2,542
C	wx	717
$+$	$+$	739
$+$	wx	2,710

Total 6,708

New combinations—21.7 per cent

The reciprocal cross, $+\ wx\ ♀ \times \dfrac{C\ +}{+\ wx}\ ♂$, shows about the same percentage of new combinations. Similar percentages are obtained if the heterozygous plant tested has one recessive allele in each chromosome.

Linkage and F_2 Ratios.—Knowing the relative proportions of non-crossover and crossover gametes for the two parents, it is a simple matter to predict the F_2 results by means of a diagram. Simplifying the frequencies somewhat by assuming exactly 20 per cent of new combinations, the results of self-pollinating F_1 plants in which the two recessive alleles are in the same chromosome, are given in figure 29. Collecting like phenotypes, we have:

C	$+$	66
C	wx	9
$+$	$+$	9
$+$	wx	16

In an actual experiment of this type Collins recorded the following frequencies:

C	$+$	1,774
C	wx	263
$+$	$+$	279
$+$	wx	420

Expressing these frequencies on a percentage basis and comparing with the frequencies expected with 20 per cent of crossing over in both male and female gametophytes, we have:

Phenotype		Expected	Observed
C	$+$	66	65.0
C	wx	9	9.6
$+$	$+$	9	10.2
$+$	wx	16	15.3

The agreement is as close as could reasonably be expected. A similar calculation of F_2 results is possible where the dominant alleles are in separate chromosomes in the F_2. If this is done,

$$\frac{C \quad +}{+ \quad wx} \quad X \quad \frac{C \quad +}{+ \quad wx}$$

	SPERMS WITH RELATIVE FREQUENCIES			
	C + 4	C wx 1	+ + 1	+ wx 4
C + 4	C + 16	C + 4	C + 4	C + 16
C wx 1	C + 4	C wx 1	C + 1	C wx 4
+ + 1	C + 4	C + 1	+ + 1	+ + 4
+ wx 4	C + 16	C wx 4	+ + 4	+ wx 16

(EGGS WITH RELATIVE FREQUENCIES — row labels on left)

Fig. 29.—Checkerboard diagram showing the phenotypes and their relative frequencies obtained in the F_2 generation of a cross between C + and + wx maize. For simplicity exactly 20 per cent of recombination is assumed.

again assuming 20 per cent of recombination, it is found that the following relative frequencies are expected:

C +	51
C wx	24
+ +	24
+ wx	1

Actual experiments give results in agreement with these calculated frequencies.

It should be recalled that two independently segregating auto-somal gene pairs, both of which show complete dominance, are expected to give the four possible phenotypes in the ratio $9:3:3:1$. If this is expressed on a percentage basis and compared with gen-eralized cases of linkage with 20 per cent recombination, we have the following:

		Phenotypes			
		$+\ +$	$+\ b$	$a\ +$	$a\ b$
Independence	$\left(\dfrac{+}{a}\ \dfrac{+}{b}\right)$	56.25	18.75	18.75	6.25
F_2	$\left(\dfrac{+}{a}\ \dfrac{+}{b}\right)$	66	9	9	16
F_2	$\left(\dfrac{+}{a}\ \dfrac{b}{+}\right)$	51	24	24	1

In both cases of linkage it is evident that phenotypes like the parents are more frequent than would be expected with inde-pendent assortment; this, in fact, is a general rule for the detec-tion of linkage. This rule breaks down, however, in the case in which linked genes have a recombination percentage of 50; the results are then identical with those expected for gene-pairs car-ried in independent chromosome-pairs. The absence of crossing over in the male of Drosophila is used to advantage in this re-spect, since when one makes a testcross with the heterozygous male, the detection of linkage is quite independent of the fre-quency of crossing over in the female.

Recombination Frequencies from F_2 Data.—It is obvious that by doing the reverse of what was done above it is possible to calculate the recombination percentage for two linked genes from F_2 zygote frequencies. The more efficient calculations of this type are rather involved and for this reason are not given in this book. Those who have occasion to calculate recombination values from F_2 data are referred to treatments of this subject by Immer and by Mather (see references). Calculation of recombination percentages from data from testcrosses to the double recessive are more direct than are those from F_2 data. Furthermore they are more efficient, *i.e.,* a value based on 100 individuals in a testcross is less subject to errors of sampling than is a value based on the same number of individuals in F_2. Calculations from F_2 data

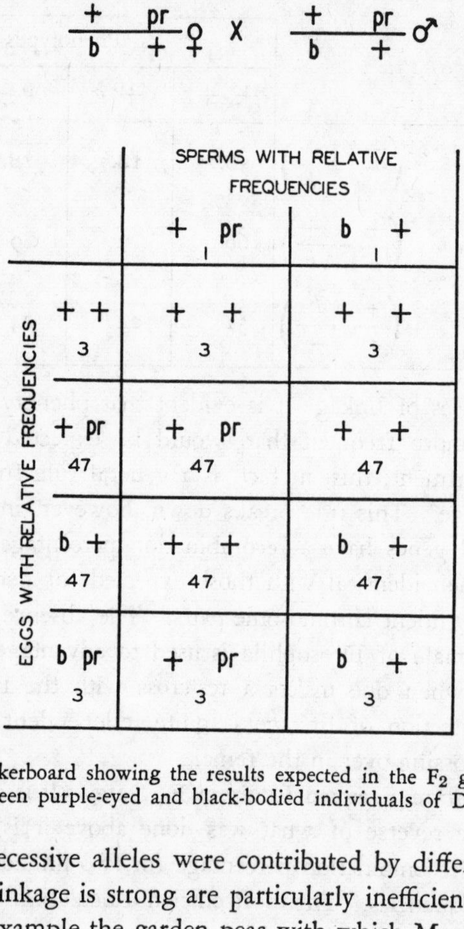

$$\frac{+}{b} \; \frac{pr}{+} \; ♀ \quad X \quad \frac{+}{b} \; \frac{pr}{+} \; ♂$$

	SPERMS WITH RELATIVE FREQUENCIES	
EGGS WITH RELATIVE FREQUENCIES	+ pr $\quad$ 1	b + $\quad$ 1
+ + $\quad$ 3	+ + $\quad$ 3	+ + $\quad$ 3
+ pr $\quad$ 47	+ pr $\quad$ 47	+ + $\quad$ 47
b + $\quad$ 47	+ + $\quad$ 47	b + $\quad$ 47
b pr $\quad$ 3	+ pr $\quad$ 3	b + $\quad$ 3

Fig. 30.—Checkerboard showing the results expected in the F_2 generation of a cross between purple-eyed and black-bodied individuals of Drosophila.

where the recessive alleles were contributed by different parents and where linkage is strong are particularly inefficient. In many plants, for example the garden peas with which Mendel worked, it is technically difficult to make testcrosses on a large scale (few seeds per pollination) and linkage studies must then be based to a large extent on segregations in F_2.

Linkage and F_2 Ratios in Drosophila.—Because of the absence of crossing over in the male of Drosophila, F_2 segregations of linked genes represent a special case. The results can, of course, be predicted just as they were in corn. As an example, let us consider the F_2 generation of the cross of a black (b) bodied female with a purple (pr) eyed male. These two genes give about 6 per cent of recombinations. The expected results are given in figure 30. Collecting like phenotypes we get:

+	+	100
+	pr	50
b	+	50
b	pr	0

or a 2:1:1:0 ratio where a 9:3:3:1 ratio would have been expected with independent inheritance. The 2:1:1:0 ratio is independent of the frequency of recombination as can be seen by study of the above diagram or, empirically, by trying various gametic frequencies for the female. In an actual F_2 from an F_1 in which the recessive alleles were in separate chromosomes, Bridges obtained the following results:

+	+	684
+	pr	300
b	+	371
b	pr	0

This is in reasonable agreement with the expected 2:1:1:0 ratio.

From the F_2 obtained by crossing F_1 individuals in which recessive alleles b and pr are in the same chromosome we would expect the results given in figure 31. Summarizing these results, we have:

+	+	147
+	pr	3
b	+	3
b	pr	47

It is clear that here the relative frequencies of F_2 zygotes are dependent on the frequency of crossing over, and, furthermore, that the calculation of the percentage of recombinations from F_2 data is relatively simple.   Such data are rarely used in Drosophila, however, since if double recessive flies are available to make the original mating, they are usually available at the proper time to make a testcross.   Clearly the testcross is the more efficient.

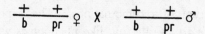

	SPERMS WITH RELATIVE FREQUENCIES	
	+ + 1	b pr 1
+ + 47	+ + 47	+ + 47
+ pr 3	+ + 3	+ pr 3
b + 3	+ + 3	b + 3
b pr 47	+ + 47	b pr 47

EGGS WITH RELATIVE FREQUENCIES

Fig. 31.—Chart showing the F_2 results of a cross between wild-type and black purple individuals of Drosophila.

Linkage Groups.—We have considered three linked autosomal genes in Drosophila: black, purple, and vestigial.   Three sex-linked genes have been mentioned: yellow, white, and bar.   These are all carried in the X chromosome.   Actually, several hundred genes are known in Drosophila and these have been shown to fall into four and only four groups, known as *linkage groups*. These, of course, correspond to the four chromosome pairs.   In another species of Drosophila, D. pseudoobscura, there are five instead of four pairs of chromosomes and, corresponding with these, there are five linkage groups. In still another species, D. virilis, there are six pairs of chromosomes and, although fewer genes are known here than in D. melanogaster, these genes fall into six linkage groups as expected. In maize, more than a hundred genes make up ten linkage groups corresponding to the ten chromosome pairs. Still other examples might be mentioned; in all forms in which adequate evidence is available, the number of linkage groups is exactly the same as the haploid number of chromosomes.

Linkage in Other Organisms.—Linkage with crossing over is known in many organisms, both plant and animal. In fact, no organism has been found in which these phenomena are known to be absent. In man there are the two well-analyzed sex-linked characters already mentioned, color-blindness and hemophilia. These must both be carried in the X chromosome and should be linked. Because both of these characters are relatively infrequent, few cases have been found in which a woman has been known to be heterozygous for both of them. Extensive searches, however, have resulted in the discovery of a few such women. Classifications of their sons have indicated that the recombination frequency for these two genes is of the order of five per cent (Haldane and Bell). In addition to a group of sex-linked genes in man, there are, of course, many autosomal genes and these we should expect to fall into twenty-three pairs of autosomes. Future work will, without question, lead to the discovery of more cases of linkage in man, but because of the difficulties of study, it may be predicted that the accumulation of data necessary to establish these will be a long-time process.

REFERENCES

Darlington, C. D. 1937. Recent Advances in Cytology. 671 pp. Blakiston's Sons and Co., Philadelphia.

Immer, F. R. 1930. Formulae and tables for calculating linkage intensities. Genetics, 15: 81–98.

Mather, K. 1938. The measurement of linkage in heredity. 132 pp. Methuen and Co., London.

Sharp, L. W. 1934. Introduction to cytology. 567 pp. McGraw-Hill, New York.

Wilson, E. B. 1925. The cell in development and heredity. 1,232 pp. The Macmillan Co., New York.

(See also references for Chapter VI.)

PROBLEMS

1. Dwarf plant (d) and pubescent fruit (p) in tomatoes are recessives to tall plant and smooth fruit. F_1 plants from a cross dwarf pubescent and tall smooth were tested by crossing to dwarf pubescent plants and the following data obtained (from Lindstrom):

   dwarf smooth 5
   dwarf pubescent 118
   tall pubescent 5
   tall smooth 161

   Using symbols instead of names, classify these data to show parental and non-parental combinations. Is there evidence of linkage? If so, what is the percentage of recombinations?

2. Red pericarp (P) in maize (the character familiar to many in red ears of corn) is a dominant character. Tassel-seed-2 (ts_2—seeds instead of stamens in the tassel) is recessive. Crosses between plants of a certain culture and colorless pericarp tassel-seed plants gave the following phenotypes (from Emerson):

   red pericarp, tassel-seed 15
   colorless pericarp, normal tassel .. 14
   colorless pericarp, tassel-seed 1,174
   red pericarp, normal tassel 1,219

   What was the genetic constitution and phenotype of the plants that gave rise to those tabulated, *i.e.*, those in the "certain culture"? What percentage of recombination would you calculate for the two gene pairs concerned?

3. Brachytic (br) is a recessive dwarf type in maize. Testcrosses between $\dfrac{P \quad +}{+ \quad br}$ and $+ \ br$ gave (from Emerson):

P	$+$	204
P	br	153
$+$	$+$	154
$+$	br	165

From a purely statistical standpoint would you conclude that P and br are linked, or are inherited independently? What is the probability that the observed ratio of parental to new combinations represents a chance deviation from the 1:1 ratio expected with independent assortment?

4. From the two crosses, $\dfrac{+\quad +}{pr\quad vg}$ ♀ $\times$ $pr\ vg$ ♂ and $\dfrac{+\quad vg}{pr\quad +}$ ♀ $\times$ $pr\ vg$ ♂, given in the text, the recombination values found were 10.7 and 13.0 respectively. Is the difference between these two values, 2.3 per cent, statistically significant? Refer to text for any additional data required.

5. What frequencies of phenotypes would have been expected in F_2 if the F_1 plants of problem 1 had been self-pollinated instead of crossed with $d\ p$ plants? What would be the expected frequencies of phenotypes in an F_2 from self-pollination if the d and p genes were carried in different chromosome pairs?

6. In Drosophila, spineless (ss) is a recessive character in the third chromosome pair. The bristles are shorter than normal. Black body color and purple eye color are linked and give about 6 per cent recombination. They are carried in the second chromosome pair. What relative frequencies would you expect for the various phenotypes among the offspring of the cross

$\dfrac{+\quad +}{b\quad pr}\dfrac{+}{ss}$ ♀ $\times$ $b\ pr\ ss$ ♂? From the reciprocal of this cross?

7. The two genes C and a_2+ in maize are complementary genes for aleurone color and are carried in separate chromosome pairs. When both are segregating, a 9:7 ratio of kernels with purple and colorless aleurone is expected in F_2. As stated in the text, C and waxy (wx) are linked with 20 per cent recombination. What phenotypic ratio would be expected in F_2 on self-pollinating plants of the constitution $\dfrac{C\quad +}{+\quad wx}\dfrac{+}{a_2}$?

8. With two pure stocks of Drosophila, one yellow (y) and the other white (w), outline the method you would use for making a homozygous stock recessive for both of these sex-linked genes. Refer to the text for the frequency of recombination between these two loci. (Note: It is possible to hold individuals of either sex of Drosophila over and mate them with individuals of the next generation. Males can be used in many matings but females cannot be used efficiently in more than one mating.)

9. Starting with two stocks of Drosophila, one purple, and one ves-

tigial, outline the crosses required to obtain a homozygous purple vestigial stock. Remember that there is no crossing over in the male of Drosophila.

10. The gene *pr* differentiates red from purple aleurone in maize, but this difference cannot be detected in kernels with colorless aleurone. The genes a_2 and *pr* are both in chromosome pair 6 and show 30 per cent recombination. What phenotypic ratio would you expect on selfing a plant of the constitution $\dfrac{+}{a_2} \dfrac{+}{pr} \dfrac{C}{+}$?

CHAPTER VI

CHROMOSOME MAPS

A TESTCROSS in Drosophila involving the two sex-linked characters scute (*sc,* scutellar and certain other bristles missing—Fig. 21) and echinus (*ec,* rough eye) gave a crossover value of 8.9 per cent. In another experiment the crossover value between echinus and crossveinless (*cv,* a wing character—Fig. 21) was found to be 9.7 per cent. It is not necessary to determine these values from separate experiments; all three loci can be followed in a single testcross. As an example, the cross $\dfrac{+\ ec\ +}{sc\ +\ cv} \times sc\ ec\ cv$ gave the following phenotypes:[1]

+ *ec* +	810
sc + *cv*	828
+ + *cv*	88
sc ec +	62
+ *ec cv*	103
sc + +	89
+ + +	0
sc ec cv	0
Total	1,980

Gene Sequence.—One striking result of this experiment is that there are only six phenotypes present, where of course eight would be expected with independent segregation. The + *ec* + and *sc* + *cv* phenotypes are like the parents of the triple heterozygote, *i.e.,* are non-crossovers. Two of the remaining phenotypes, + + *cv* and *sc ec* +, can be accounted for by crossing

[1] The examples of Drosophila crosses used in this chapter are from the work of Bridges and Olbrycht.

over between *sc* and *ec*. The other two, $+$ *ec cv* and *sc* $+$ $+$, can be accounted for by assuming crossing over between *ec* and *cv*. This experiment tells us that the three genes *sc, ec,* and *cv* are linked, but an even more interesting result is that we cannot interpret the results in any simple way without assuming that *the genes are arranged in the chromosome in the sequence sc-ec-cv* (or *cv-ec-sc*). In other words, this type of testcross tells us the order of the genes in the chromosomes.

Gene Loci.—The relative constancy of crossover values and the constant order of genes in chromosomes imply that every gene occupies a fixed position in a chromosome, and its allele a corresponding position in a homologous chromosome. This will be demonstrated in a more direct way later. Such a position is known as a *locus* (plural, *loci*). A particular locus is designated by the name of the mutant allele that occupies the locus. We speak of the "white" locus or the "yellow" locus, for convenience designated by gene symbols—*i.e.,* the *w* locus.

Linear Arrangement and Chromosomes.—The linear arrangement of genes in chromosomes and the fact that this arrangement can be demonstrated by studying crossing over is not surprising, when we remember the fact mentioned in the previous chapter, that chromosomes are long and thread-like at the stage at which crossing over occurs (pachytene). We do not, of course, imply by linear arrangement a straight line, but rather that the genes are arranged in a manner similar to beads strung on a loose string.

Crossing Over and Physical Distances.—Returning to the example of the cross in which the scute, echinus and crossveinless characters were followed, it can be seen that the crossover percentage for two loci must be related to the distance between the loci. If *ec* is between *sc* and *cv,* then the distance between *sc* and *cv* must be greater than either that between *sc* and *ec* or that between *ec* and *cv*. The crossover percentage for *sc* and *cv* is exactly the sum of the values for *sc-ec* and *ec-cv,* since all recorded crossovers, whether between *sc* and *ec* or between *ec* and *cv,* are crossovers between *sc* and *cv*. If chiasmata were distributed entirely at random along the length of a chromosome pair, then of course it would follow that crossover values would be related to physical distances in a simple linear way. Actually the relation is not as simple as this; chiasmata are not distributed purely at random. This question will be considered in more detail later, but it can be said here that certain sections of a chromosome pair show more crossing over per unit physical length than do other sections of the same chromosomes.

An Analogy.—The relation of crossing over to distances and linear order has been illustrated by an analogy with railroad time tables. If we consider the times at which a given train arrives at three specified stations, these times will tell us, without any question, the order in which these stations are arranged along the right of way. This is equivalent to determining that three loci are arranged in a particular order in the chromosomes. Knowing the train time for a run between stations, we can deduce roughly the distance apart of these stations. For a given train there will be a fairly close correlation between time and distance traveled. But of course this will not be exact—a 50-mile run through mountainous country may take twice as long as the same length run over level country. This is comparable to regional differences in crossing over. Even the time between two given stations is not constant; it will vary depending on whether a freight train or streamliner is making the run. This illustrates another analogy. If we were to set out to determine the order of three stations, not too far apart, by deducing distances from times required to make the runs between various combinations of two of them, we would arrive at the correct result if we based our deduction on times by a given train. Thus, suppose the following information is available:

a to c .... 2 hours
b to c .... 1 hour
a to b .... 3 hours

It is clear that if the train runs at a reasonably constant speed, the sequence of stations must be a-c-b. But if times were taken from runs by different trains with no knowledge of their speeds the above order might very well be incorrect. If the a to b and b to c times were taken from a freight train schedule where the speed was 50 miles per hour and time a to c was taken from a streamliner schedule, with the train averaging 100 miles per hour, the order would of course be a-b-c. In the same way using separate two-point measurements of crossover values, it is possible to arrive at an erroneous deduction as to the order of genes.

Crossover Regions.—In a testcross such as the one given above, the intervals between adjacent loci being studied are referred to as *crossover regions*, or sometimes simply as regions. It is conventional to number these regions from left to right. Thus, in the above example, there are 150 crossovers in region 1 (between *sc* and *ec*) among a total of 1,980 individuals or 7.6 per cent. In region 2 (between *ec* and *cv*) there are 192 or 9.7 per cent of crossovers.

Different Arrangements of Linked Genes.—With three linked genes heterozygous there are four arrangements possible, *i.e.*, $\dfrac{a \quad b \quad c}{+ + +}$, $\dfrac{a \quad b \;+}{+ + c}$, $\dfrac{a \;+\; c}{+ \; b \;+}$, and $\dfrac{+\; b \quad c}{a \;+\; +}$. It makes no difference which of these is used in making a testcross involving all three genes (technically a *three-point-test*). For example, a testcross, quite independent of the one given above, involving scute, echinus, and crossveinless was made in the following way:

$$\frac{+ \; + \; +}{sc \;\; ec \;\; cv} \times sc \;\; ec \;\; cv$$

The results were:

Non-crossovers	$\dfrac{+ \; + \; +}{sc \;\; ec \;\; cv}$	352
		284
Crossovers region 1 (8.9 per cent)	$\dfrac{+ \;\; ec \;\; cv}{sc \;\; + \; +}$	26
		46
Crossovers region 2 (12.4 per cent)	$\dfrac{+ \; + \;\; cv}{sc \;\; ec \;\; +}$	38
		62

The crossover values here are somewhat higher than those obtained in the previous cross but they are still reasonably close.

The crossover value for *sc-cv*, 8.9 + 12.4, or 21.3, is, of course, likewise greater than that obtained previously.

Chromosome Maps.—With a method of determining the order of loci in a chromosome and a measure of distance, we can readily construct a chart showing the spacing of genes in the chromosome. Such a chart is known as a *chromosome map*. In order to do this we must have a *map unit* to represent spacing. The standard unit used for this purpose is arbitrarily taken as the distance that will give, on the average and under "standard" environmental conditions, one crossover per hundred gametes. In other words, two genes that show 10 per cent of crossing over, are 10 units apart. Applying this to the first three point test given above involving *sc, ec,* and *cv* we could make the following map:

sc	7.6	*ec*	9.7	*cv*

Double Crossing Over.—Still more loci can be added to this map. Let us consider two additional sex-linked genes in Drosophila, cut (*ct*, a wing character—Fig. 21) and vermilion (*v*, an eye color). Using these in combination with one of those previously used, crossveinless, we can make another three-point testcross. We may consider the following one of the four possible types of heterozygous females:

$$\frac{+ \quad ct \quad +}{cv \quad + \quad v} \times cv \; ct \; v.$$

The types obtained in one such test were:

Non-crossovers	$\dfrac{+ \;\; ct \;\; +}{cv \;\; + \;\; v}$	759 766
Crossovers region 1 (7.7 per cent)	$\dfrac{+ \;\; + \;\; v}{cv \;\; ct \;\; +}$	73 80
Crossovers region 2 (15.0 per cent)	$\dfrac{+ \;\; ct \;\; v}{cv \;\; + \;\; +}$	140 158
(0.2 per cent)	$\dfrac{+ \;\; + \;\; +}{cv \;\; ct \;\; v}$	2 2

These results differ from those in previous examples in one important respect. There are two phenotypes among the offspring that are not accounted for by single crossing over in either region 1 or region 2. They can be accounted for, however, by assuming that *double crossovers occur;* that is, instances in which *a crossover in region 1 is accompanied by a crossover in region 2.* As a matter of simple probability these would be expected to be relatively infrequent. Inspection of the data shows that this is the case. Doubles may be expected in any three-point test in which the total distance covered by the three loci is more than 10 to 20 map units. This point will receive further consideration later. Neglecting sampling errors, double crossovers are always less frequent than either of the component single crossovers. If we consider only the *cv* and *ct* loci, it is evident that the double crossover classes will have to be counted with the singles in region 1 to get the total crossing over in region 1. This will be

$$\frac{73 + 80 + 2 + 2}{1,980} \times 100,$$ or 7.9 per cent. In the same way, considering only region 2, it is clear that the double crossovers will have to be included to get the total crossing over for this region. In other words, since the double crossovers are crossovers in both regions 1 and 2, they must be counted as crossovers twice, once for each region. The crossover value for region 2 is 15.2 per cent.

Crossover and Recombination Frequencies.—If we consider the map distance between *cv* and *v*, we see that it is the sum of the *cv-ct* and the *ct-v* distances or 23.1 units. But, if the test had been made omitting the *ct* locus, we would not have obtained this value. The value 23.1 includes the double crossover classes twice, once in each region, and without following the middle locus these would not have been detected. This can be illustrated by ignoring the classification for the cut character in the example. The + + + and *cv ct v* phenotypes would then appear as parental types, *i.e.,* as + + and *cv v*. The value for *cv* to *v* would be 22.7 instead of the 23.1 obtained by adding the crossover values for the two component regions. The difference is, of course, small and in this particular case we would not need to worry about it. For larger intervals, however, the relative proportion of double crossover types increases and the error becomes much more important. If we consider only the *cv* and *v* loci in the above test, the value of 22.7 is not, therefore, the true crossover value. It is too small because of undetected double crossovers. Such a value is known technically as a *recombination value* as contrasted to the true crossover value. In building up maps, it is necessary to work with regions so short that no doubles occur, if true crossover values are desired. This can be determined experimentally by following a locus within the region, *i.e.*, breaking the one region into two. For example, for the *ct-v* region we can add the lozenge (*lz*, eye shape) locus in a test of a $\dfrac{ct \quad v}{lz}$ female. Actually, such a test would show no double crossovers. It is not always possible to work with intervals short enough to give true crossover values, either because the number of mutant characters is not sufficient, or because they are not distributed properly. In such cases we must be content with an approximation until more loci can be used. In the examples of Drosophila crosses used in this chapter, the regions are short enough to give true crossover values.

Rule for Determining Gene Sequence.—We are now in a position to give another and simpler rule for determining the order of genes from a three-point test. The double crossover classes have the two alleles of the middle gene exchanged with respect to those of the other two. Since this requires two crossovers simultaneously, the classes resulting will be the least frequent pair of phenotypes or, as in the *sc-ec-cv* example, will be entirely absent. Reversing the argument, it is clear that, if we take the least frequent (or absent) pair of *contrary crossover* classes (complementary products of crossing over) and determine by inspection which pair of alleles in the F_1 heterozygote must be exchanged to get these classes, these alleles will represent the middle locus.

Returning to the *cv-ct-v* three-point test, we have the following map distances: *cv-ct*, 7.9 and *ct-v*, 15.2. With this information we cannot put the genes *ct* and *v* on the *sc-ec-cv* map, because we do not know on which side of *cv* they lie. We can determine this, however, by making a third three-point test, for example, one involving *ec, cv,* and *ct*. A test of this kind follows:

$$\frac{+ \quad cv \quad +}{ec \quad + \quad ct} \times ec\ cv\ ct$$

	+ *cv* +	2,207
	ec + *ct*	2,125
(10.1 per cent)	+ + *ct*	265
	ec cv +	273
(8.3 per cent)	+ *cv ct*	223
	ec + +	217
(0.2 per cent)	+ + +	5
	ec cv ct	3

The last pair of contrary crossover phenotypes are least frequent and these can be obtained from the test heterozygote by exchanging the two alleles of the crossveinless gene. It follows, therefore, that *cv* is the middle locus and that the order is *ec-cv-ct*.

Sequence of Many Genes.—Combining the results of the separate three-point tests, we see that the order of the five sex-linked genes involved must be *sc-ec-cv-ct-v*. As has been pointed out, we could just as well write these in the reverse order. For the sake of uniformity, it is customary in making chromosome maps to arbitrarily decide which end of a sequence of genes is to be written to the left. In the X chromosome of Drosophila, it has been decided to write *sc* to the left of *ec*. Positions are then referred to as to the "right" or "left" of a point of reference.

Additional genes can be added to a map by the principles illustrated above. Using the character garnet (*g*, a pinkish eye color), results of a cross $\frac{+ \quad v \quad +}{ct \quad + \quad g} \times ct\ v\ g$ are as follows:

Crossover region	Phenotype	Number of individuals
0	$+$ v $+$	1,588
	ct $+$ g	1,485
1 (13.5 per cent)	$+$ $+$ g	277
	ct v $+$	266
2 (9.9 per cent)	$+$ v g	184
	ct $+$ $+$	214
1, 2 (0.3 per cent)	$+$ $+$ $+$	5
	ct v g	8
	Total	4,027

It will be noted that here there are two genes involving eye color. There is no difficulty here since, for these two segregating gene pairs, four phenotypes can be distinguished easily. Some combinations of characters cannot be distinguished in this way. For example, it would be quite inconvenient to use the white eye character in the same cross with vermilion eye, since the vermilion character cannot be classified in the white eyed flies.

The above results give the order ct-v-g and the garnet locus can be added to the right of the five already considered, to give the order sc-ec-cv-ct-v-g. Calculating the distances, we can construct the map:

sc	ec		cv	ct		v	g
7.6	13.7		7.9	15.2		10.2	
	10.3		8.5	13.8			

In three of the regions we have put down crossover values obtained from two separate experiments. The differences between the pairs of values illustrate the variation one is likely to get in actual experiments of this type.

Any number of loci can be added to a chromosome map, either at the end or between two loci already located. The testcross $\frac{+\ \ +\ \ +}{v\ \ m\ \ g} \times v\ m\ g$ would show that the m (miniature wing) locus lies between v and g.

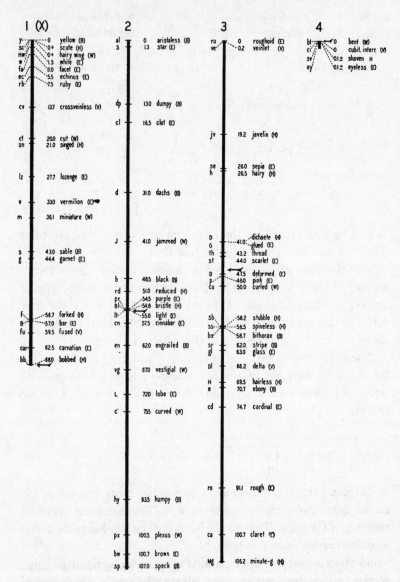

Fig. 32.—Linkage maps of the four chromosomes of Drosophila melanogaster. The letters in parentheses indicate the part of the fly affected by the mutant character concerned: B, body; E, eyes; H, hairs or bristles; V, venation of wings; W, wings. Positions of centromeres (spindle attachments) are indicated by arrows. (After Bridges.)

Methods of Constructing Chromosome Maps.—Several hundred genes have been "mapped" in Drosophila. Figure 32 shows such a map with only the loci most often used indicated. The maps of the autosomes are made in exactly the same way as are those of the X chromosome. The X chromosome is easier to work with, since to make the equivalent of a three-point testcross, a triple recessive stock is not needed—heterozygous females may be mated to males of any constitution, and only male offspring used for determining recombination values. Similar maps have been made for other species of Drosophila, for other animals, and for several plants. Figure 33 shows the chromosome maps of maize.

It will be noted that in all of these maps loci are assigned positions that are designated with a number obtained by adding the crossover values for all known intervals to the left. Thus the *v* locus is at 33 and *g* is at 44.4. The map distance between them is 11.4. As we have already pointed out, one would not expect, in testcrosses involving *v* and *g,* to get exactly this crossover value. The map positions are based on the averages of many carefully made tests.

Variability.—If the measurement of crossover and recombination percentages were subject only to errors of sampling, there would be no difficulty in constructing chromosome maps from two-point linkage data. Because of variability due to other factors, however, one may be led to erroneous conclusions in attempting to determine gene sequences in this way. This may be illustrated by an actual example in maize involving the three fourth chromosome characters sugary endosperm (*su*), Tunicate (*Tu*—long glumes—"pod corn"), and a glossy seedling character (*gl₃*). As indicated by the map in figure 33, the standard recombination value for *su* and *Tu* is 29 per cent. This region does not include loci that can be marked with workable characters; presumably the value 29 is too low. This, however, need not concern us here. A two-point test of *su* and *gl* gave a recombination percentage of 23 and a test involving *Tu* and *gl* gave a value of 11 per cent. These results suggest that the *su-Tu* distance is greater than that between *su* and *gl* or that between *Tu* and *gl*. This

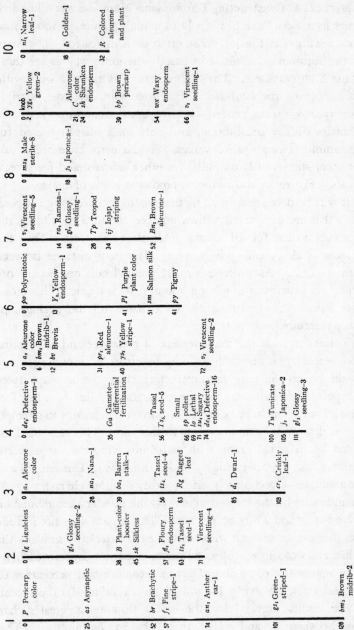

Fig. 33.—Linkage maps of the ten chromosomes of maize. (From Emerson, Beadle, and Fraser.)

would of course indicate the order *su-gl-Tu*. The three-point test

$$\frac{+\quad Tu\quad +}{su\quad +\quad gl} \times su + gl$$ gave the following results (from Emerson, Beadle, and Fraser):

Crossover regions					Number of individuals
0		1	2	1, 2	
$\begin{array}{cc}+ & su\\ Tu & +\\ + & gl\end{array}$		$\begin{array}{cc}+ & su\\ + & Tu\\ gl & +\end{array}$	$\begin{array}{cc}+ & su\\ Tu & +\\ gl & +\end{array}$	$\begin{array}{cc}+ & su\\ + & Tu\\ + & gl\end{array}$	
291	206	133 91 (27.5%)	39 37 (9.5%)	8 8 (2.0%)	813

This test shows conclusively that the order is not *su-gl-Tu,* but *su-Tu-gl.* In this case the results from two-point tests were quite misleading.

Maps from F$_2$ Data.—We pointed out in the preceding chapter that it is possible to calculate linkage values from F$_2$ data. Maps can be constructed on the basis of such values. It is also possible to use F$_2$ results involving the segregation of more than two linked gene-pairs. The difficulties encountered with two linked gene-pairs are multiplied as more gene-pairs are involved, so that in material where testcrosses are readily made, F$_2$ results are very seldom used in constructing maps.

MULTIPLE CHIASMATA

The relation between single chiasmata and crossovers was considered in Chapter V. We saw that the proportion of tetrads in which a chiasma occurs between two specified loci is twice that of single chromatids that show a crossover between the same two loci. If we wanted to make a chromosome map by defining a unit of distance as that within which an exchange (or a chiasma cytologically) would fall on the average once in 100 tetrads we would come out with all values multiplied by two. The relations are just as simple when we consider the relation between double crossovers and the double exchanges that give rise to them, but they are not quite as obvious.

Types of Double Exchanges.—Since the four chromatids present at the time of crossing over are genetically of two kinds only (Fig. 34), it makes no difference which two are involved in a single exchange so long as these are not *sister chromatids* (the two derived from one during the preceding division of pachytene threads). As a matter of fact, there is evidence that sister chromatids do not exchange segments as do non-sister chromatids, but this evidence is indirect and complicated, and will not be presented in this book. The following discussion assumes, then, that such *sister chromatid crossing over* never occurs. Any one of the four possible single exchanges (Fig. 34) is obviously equivalent

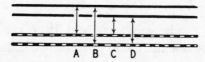

A   B   C   D

Fig. 34.—Diagram of the four possible single non-sister-strand exchanges. The two chromatids taking part in an exchange are indicated with arrows. Disregarding all other exchanges, A, B, C and D are equivalent; each gives two non-crossover and two crossover chromatids.

to any other, *i.e.,* each results in two crossover and two non-crossover chromatids. In the case of double exchanges, on the other hand, it does make a difference which two chromatids are involved at the second of the two component single exchanges. This is illustrated in figure 35. Here the exchanges represented at the right involve four different pairs of chromatids with reference to those involved in the exchange shown at the left. A convenient way of designating these types is according to the number of chromatids involved in crossovers, considering both exchanges. Thus a 2-strand (strand is equivalent to chromatid here) double exchange gives rise to two non-crossover chromatids and two double crossover chromatids. The two types of 3-strand doubles are equivalent to each other; the four resulting chromatids are one non-crossover, one single crossover in region 1, one single crossover in region 2 and one double crossover. The four-strand exchange gives all single crossover chromatids, two for each region. There is evidence of an indirect nature that, in Drosophila, the four possible types of double exchanges are equally frequent; in other words, that 2-, 3- and 4-strand double exchanges occur in the ratio of 1:2:1.

Chromatids Recovered Following Double Exchanges.—It will be seen that a single chromatid recovered from a tetrad in which

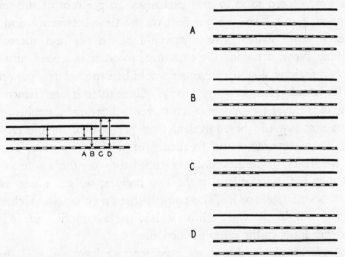

Fig. 35.—Diagram of the relations of two separate exchanges. With the left exchange fixed as indicated there are four possible double exchanges, as indicated. A is a 2-strand, B and C are 3-strand, and D is a 4-strand exchange. These are designated by the number of chromatids involved in the two exchanges.

two chiasmata occurred need not necessarily be a double cross-over. In fact, if we collect the various types of chromatids resulting from double exchanges we see that there are four types as follows:

Non-crossovers 4
Single crossovers, region 1 4
Single crossovers, region 2 4
Double crossovers 4

The four types occur with equal frequencies. This means that, where a single product of a given meiotic division is recovered, as in the eggs of animals or the megaspores of plants, or where the four products must be considered without relation to one

another as in animal sperms or plant microspores, a double crossover product is recovered only one fourth of the time following a
double exchange. If, then, we want to calculate, from the frequency of double crossover gametes, the frequency of double exchange tetrads, we simply multiply by four. To illustrate these
relations let us consider an hypothetical example with the three
loci *a-b-c* spaced so as to give exchanges 20 per cent of the time
between *a* and *b* and 20 per cent of the time between *b* and *c*.
The crossover values obviously would be 10 per cent for each
region. Now, assuming the exchanges to occur independently in
the two regions, double exchanges would be expected in 4 per cent
of the tetrads ($0.2 \times 0.2 \times 100$). Since there is one chance in
four of recovering a double crossover following a double exchange, this would be equivalent to 1 per cent of double crossovers. If we predict the frequency of double crossovers from
the frequency of single crossover chromatids on the assumption
that they are independent in the two regions, we get 1 per cent
($0.10 \times 0.10 \times 100$). The point is that on either basis (chiasmata or single recovered chromatids), the assumption of independence leads to the same prediction.

   Interference.—In the above argument we have assumed that
double exchanges occur independently in two adjacent regions.
This assumption is incorrect as can be seen from the examples
given earlier in this chapter. In the first three-point test involving *sc, ec,* and *cv* (page 93) there was 7.6 per cent of crossing
over in region 1 and 9.7 per cent in region 2. There were no
double crossovers at all in this cross where, were crossing over
in the two regions independent, we would expect 0.7 per cent.
This can only mean that one exchange interferes with an exchange in adjacent sections of the chromosome pair. This phenomenon is known as *interference.* This is usually expressed in
a somewhat different way, namely by a *coefficient of coincidence.*
This is obtained by dividing the actual proportion of doubles obtained by the proportion expected with complete independence; in

the cross illustrated, $\dfrac{0}{.007} = 0$. A coincidence value of 0 means that interference is complete. A value of 1 means no interference, or independence of crossing over for the two regions under consideration. In the example of the three-point test in maize, given on page 105, the recombination values for the two regions are 29.5 and 11.5 per cent. There are 2 per cent of double crossovers where 3.39 per cent would be expected with independent crossing over in the two regions. The coincidence value is therefore 0.59 (2 ÷ 3.39). Other examples could be given in which the coincidence varies from 0, as in the first example, to 1.0. It can be seen from the relations between exchanges and detected crossovers worked out above that the result is the same whether coincidence is calculated on the basis of tetrad exchange frequencies or on the basis of single chromatid crossover frequencies. In practice, since data are most often obtained in the form of crossover frequencies, this type of calculation is usually more convenient.

Multiple Exchanges.—If more than three loci are followed in a testcross, as is frequently the case, triple or even quadruple crossovers may be recovered. A triple crossover is counted in calculating the crossover or recombination percentage for each of the three regions involved; hence it is counted three times in calculating map units. The same principle applies to quadruple crossovers.

It is a simple matter to work out the relations between triple and quadruple exchanges and triple and quadruple crossovers. If this is done the following relations can be put down:

Exchange	Recovered single chromatids with relative frequencies				
	Non-crossovers	Singles	Doubles	Triples	Quadruples
none	1				
1	1	1			
2	1	2	1		
3	1	3	3	1	
4	1	4	6	4	1

Limit of Recombination Frequencies.—Since for terminal loci, only odd numbered crossovers will give recombinations for heterozygous genes at these loci, it is clear from the above relations that, no matter what distribution of single or multiple exchanges we have between the two loci, there will never be more than 50 per cent of recombination; the non-crossover plus the even numbered crossover chromatids are equal to the odd numbered crossover chromatids for all multiple exchanges. Thus if we take two loci 100 map units apart they will give no more than 50 per cent of recombination (except, of course, for errors of sampling). The fact that maps may be more than 50 units long is of course accounted for by the fact that they are built up of short regions within which no double crossovers occur. Theoretically there is no limit to the length of a chromosome in terms of map units; in practice, no map 200 units long has yet been established, though about 180 units are known in a few cases.

REFERENCES

Bridges, C. B., and T. H. Morgan. 1919. The second chromosome group of mutant characters. Carnegie Inst. Washington, publ., 278: 123–304.

Bridges, C. B., and T. H. Morgan. 1923. The third chromosome group of mutant characters of Drosophila melanogaster. Carnegie Inst. Washington, publ., 327. 251 pp.

Emerson, R. A., G. W. Beadle, and A. C. Fraser. 1935. A summary of linkage studies in maize. Memoir 180, Cornell Univ. Agric. Exper. Sta. 83 pp.

Mather, K. 1938. Crossing over. Biol. Reviews, 13:252–292.

Matsuura, N. 1933. A bibliographical monograph on plant genetics. Hokkaido University, Sapporo.

Morgan, T. H., C. B. Bridges, and A. H. Sturtevant. 1925. The genetics of Drosophila. Bibliogr. Genetica, 2:1–262.

Stern, C. 1933. Faktorenkoppelung und Faktoren-austausch. Handbuch Vererbungswiss. 1, H, 331 pp.

PROBLEMS

(Note: The data given in problems 1, 4, and 6 in this chapter are from actual examples of three-point tests in maize. Purely arbitrary symbols are substituted for the ones usually used.)

1. Determine the constitution of the heterozygous parent, the gene sequence, and the recombination values from the following data:

+	b	+	104
a	b	c	180
a	+	c	109
+	+	c	21
a	+	+	5
+	b	c	5
a	b	+	31
+	+	+	191

2. If *a, b,* and *c* are linked and in the order given, with *a* and *b* 20 map units apart (assume that no doubles occur within this region) and *b* and *c* 10 map units apart, what frequencies of the various phenotypes would you expect from the testcross $\frac{+ + +}{a\ b\ c} \times a\ b\ c$ with a coincidence of 1.0? With a coincidence of 0.5?

3. Three genes in the X chromosome of Drosophila are located on the chromosome map as follows: *a*—21, *b*—31, and *c*—44. Given a coincidence value of 0.2, calculate the theoretically expected frequencies of phenotypes among 1,000 individuals from the cross $\frac{+\ b\ +}{a\ +\ c}$ ♀ $\times a\ b\ c$ ♂.

4. From the five sets of data given in the following table determine the order of genes by inspection, *i.e.,* without calculating recombination values.

Phenotypes observed in 3-point testcross	Number of individuals				
	1	2	3	4	5
+ + +	317	1	30	40	305
+ + c	58	4	6	232	0
+ b +	10	31	339	84	28
+ b c	2	77	137	201	107
a + +	0	77	142	194	124
a + c	21	31	291	77	30
a b +	72	4	3	235	1
a b c	203	1	34	46	265

5. In the third example given in the above table, assume an order other than that chosen as the correct one and, on the basis of

this order, calculate the recombination values for the two regions, the percentage of doubles, and the coincidence.

6. From the data given below for the two three-point testcrosses involving (1) *a*, *b*, and *c*, and (2) *b*, *c*, and *d*, determine the sequence of the four genes *a*, *b*, *c*, and *d*, and the three single-region recombination values.

(1)				(2)			
+	+	+	669	*b*	*c*	*d*	8
a	*b*	+	139	*b*	+	+	441
a	+	+	3	*b*	+	*d*	90
+	+	*c*	121	+	*c*	*d*	376
+	*b*	*c*	2	+	+	+	14
a	+	*c*	2,280	+	+	*d*	153
a	*b*	*c*	653	+	*c*	+	64
+	*b*	+	2,215	*b*	*c*	+	141

7. With three linked genes in maize, two recessive and one dominant, and each present singly in a separate strain, outline the procedure necessary to make a three-point testcross.

8. Given three stocks of Drosophila each homozygous for a different one of the three sex-linked recessives scute (*sc*), crossveinless (*cv*), and cut (*ct*), outline the procedure you would follow for making a triple recessive stock *sc cv ct*. (Look up loci on map, figure 32, and see note in problem 8 of Chapter V.)

9. Assuming *a* and *b* to be second chromosome genes in Drosophila located 10 units apart and *c* and *d* to be third chromosome recessives 5 units apart, determine the phenotypes and their frequencies expected from the cross $\dfrac{+\ \ +}{a\ \ b}\ \dfrac{+\ \ +}{c\ \ d}$ ♀ × *a b c d* ♂.

10. Given two Drosophila stocks *y rb*, and *w*, how would you go about making a stock homozygous for *y w rb*? (Look up loci of these genes on chromosome map.)

CHAPTER VII

RELATION OF CROSSING OVER TO MEIOSIS

ATTACHED-X CHROMOSOMES

SEX-LINKAGE in Drosophila was described in Chapter II. In this type of sex-linkage the X chromosome of a son comes from the mother and one X of a daughter is contributed by the father —thus giving rise to the expression "crisscross" inheritance. With certain laboratory stocks of Drosophila these relations break down. Thus if females of a particular stock of yellow (y, the same body color gene already considered in Chapter V), are crossed to wild-type males we get the rather surprising result that the F_1 females are all yellow like the mothers and the F_1 males are all wild-type like the father. This behavior is repeated indefinitely in subsequent generations—*i.e.*, all females are yellow, and all males are wild-type. This is of course quite different from the usual type of sex-linkage.

The explanation of this peculiar deviation from the normal condition is simple. Cytological examination of the somatic chromosomes of the females of stocks that give this result shows that the two X chromosomes are *attached to each other at their centromere ends,* and in addition to these attached-X chromosomes there is a Y chromosome present in the females. Since the X chromosomes are attached, they should both go to the same pole at the first meiotic division. Actually, the two X's do behave as a single unit and regularly disjoin from the Y chromosome as indicated in figure 36. On the basis of this behavior, the inheritance from mother to daughter and from father to son becomes clear. Indicating attached-X's as $\widehat{XX}$, it is evident that in addition to $\widehat{XX}$ Y females like their mothers and normal males like their fathers, the attached-X line regularly produces two other zygotes, individuals with three X chromosomes (X͡X X) and individuals with two Y chromosomes and no X chromosome. The former usually fail to survive to the adult stage and, if they do, the individuals are phenotypically distinguishable, and are sterile; this type of individual will be discussed in connection with sex-determination. The other type, YY, dies in the egg stage. Because of the inviable or sterile zygotes, X͡X lines are only half as productive as are normal lines.

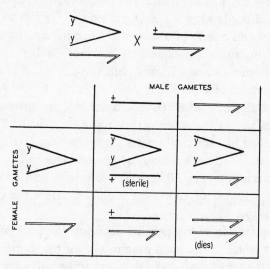

Fig. 36.—Chart showing the inheritance of the sex-linked recessive yellow body color (*y*) in a strain of Drosophila with attached-X chromosomes. The Y chromosome in this and the following diagrams is represented in outline.

Verification of Chromosome Theory.—In addition to providing the geneticist with an exceedingly useful tool, attached-X chromosomes are of interest in another connection. Such attachment of the X chromosomes was first inferred by L. V. Morgan from anomalous genetic results. The attached-X hypothesis was developed to account for these genetic results, and cytological examination showed the correctness of the hypothesis. This, then, is another example of the striking agreement between breeding results and chromosome behavior that originally led to the chromosome theory of heredity and that has repeatedly shown the correctness of that theory.

Crossing Over in Four Strand Stage.—Attached-X chromosomes provide one means of demonstrating the correctness of an assumption of which we have consistently made use in previous chapters, namely, that a given crossover involves only two of four chromatids. As first shown by Anderson, it is possible by means of special techniques to obtain attached-X females in which the

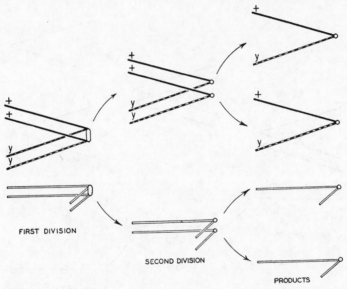

FIRST DIVISION

SECOND DIVISION

PRODUCTS

Fig. 37.—Schematic representation of the behavior of X and Y chromosomes at meiosis in attached-X females of Drosophila heterozygous for yellow, and with no crossing over between the two attached X chromosomes.

two X's differ in their gene content—*i.e.,* are heterozygous. If phenotypically wild-type X͡X females, heterozygous for yellow, are mated to males of any type, their daughters will be of two phenotypes, about 80 per cent wild-type and about 20 per cent yellow bodied. The occurrence of yellow daughters, which can be shown to have two *y* alleles and to have their X's still attached, shows that there must have been two *y* alleles present in the mother at some time during the meiotic divisions, for otherwise there would be no way for an egg to receive two such alleles. The yellow daughters are accounted for by assuming that four chromatids are present at the time crossing over occurs, and that crossing over occurs between the centromere and the locus of yellow. Figure 37 illustrates the meiotic behavior without crossing over in attached-X females. The result of divisions such as this is either an egg with a Y or an egg with attached X's, one X with y^+, the other with y. Figure 38 illustrates the

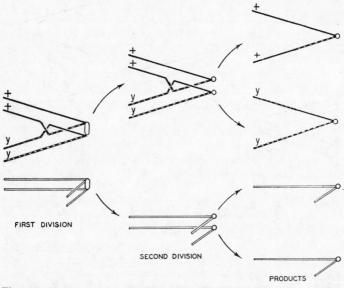

FIRST DIVISION

SECOND DIVISION

PRODUCTS

Fig. 38.—Meiosis in attached-X females of Drosophila with a single exchange between the centromere and the locus of yellow. Note that the attached-X gametes are homozygous for either yellow or its wild-type allele.

consequences of meiotic divisions during which a particular type of crossover has occurred between the centromere and the y locus. Statistically, such divisions will give eggs of the following kinds with the relative frequencies indicated:

With a Y chromosome 2
With $\widehat{XX}$, homozygous for y 1
With $\widehat{XX}$, homozygous for y^+ 1

Daughters homozygous for the y^+ allele will be phenotypically indistinguishable from heterozygous sisters, but they can of course be identified by testing them in the next generation; they give no yellow females among their progeny. In the case of a semi-dominant gene, such as bar, both homozygotes can be detected phenotypically.

Since there is no simple way of getting females homozygous for an X chromosome gene for which the attached-X mother was

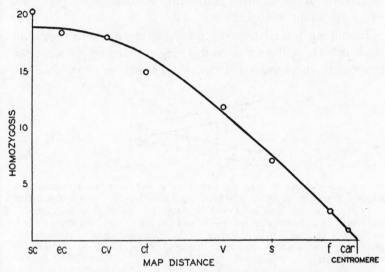

Fig. 39.—Percentage homozygosis of recessive alleles of genes heterozygous in attached-X females of the preceding generation. Circles represent observed values, while the curve represents values calculated from the observed crossover frequencies. (From Beadle and S. Emerson.)

heterozygous, except as indicated above, we conclude that crossing over does occur at a stage of meiosis at which four chromatids are present. This is often spoken of as crossing over in the *four-strand stage,* or, sometimes, in the *double-strand stage.*

Homozygosis Frequencies.—In dealing with attached-X chromosomes, the frequency with which one allele of a heterozygous gene-pair becomes homozygous among the following generation is frequently referred to as the *homozygosis frequency.* It follows from the explanation given that the homozygosis frequency should

be a function of the amount of crossing over between the centromere and the locus under consideration. When this is studied
experimentally, it is found that homozygosis values are very low
for carnation (*car*) and increase in direct proportion to map distance up to about garnet (*g*). For genes to the left of garnet
the homozygosis frequency increases, but the relation to map distance is no longer linear (Fig. 39). This non-linear relation is
a result of double crossing over. The homozygosis curve tells us
that yellow is at the distal end of the X chromosome—*i.e.*, the
end away from the centromere.

Examining the relation of crossing over and homozygosis in
more detail: of the four possible single exchanges between the
centromere and a marked locus, we see that only two result in

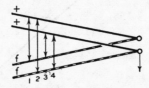

Fig. 40.—Scheme showing the four possible non-sister-strand exchanges in
attached-X females of Drosophila heterozygous for forked. On the assumption
that the pair of strands indicated with an arrow is recovered in the egg nucleus,
the results are: 1—not detected, 2—homozygous for +, 3—homozygous for f,
and 4—not detected. It is important to the understanding of subsequent diagrams that this particular method of determining the consequences of a given
type of exchange be understood.

X̂X̂ gametes homozygous for one or the other of the two alleles
present at this locus (Fig. 40). Designating the four possible
exchanges by number and indicating the pair of chromatids that
is recovered in a given egg by an arrow (Fig. 40), it is seen
that exchange 1 does not involve the recovered strands and is
therefore not detected. Exchange 4 involves both recovered
strands but, since it merely interchanges distal segments of the
two recovered strands, does not result in homozygosis. As indicated in the diagram this exchange is not detected. Exchange 2
results in a gamete homozygous for the f^+ allele and is detected
only by genotypic test. Exchange 3 gives a gamete homozygous
for the f allele and is phenotypically detected. It therefore fol-

lows that homozygosis for a particular allele of a gene is equal to one half the crossover frequency to the right of the locus of the gene, or is equal to one fourth the corresponding exchange frequency. These relations apply only for genes close enough to the centromere to justify disregarding double exchanges between the locus being studied and the centromere.

Determination of Genotypes.—If two genes, say garnet and forked, are heterozygous as indicated in figure 41, the four possible single exchanges between the two genes will give the results indicated. Exchange 1 is not detected, 2 gives a homozygote for g^+ which cannot be distinguished phenotypically from a non-

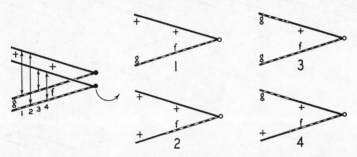

Fig. 41.—Results of the four possible non-sister-strand exchanges between the loci of garnet (g) and forked (f) in attached-X females of Drosophila. The pair of chromatids indicated by an arrow is assumed to be recovered in the egg nucleus in all cases.

crossover, 3 gives a homozygote for g, and 4 simply interchanges the two alleles of garnet with respect to their relation to the alleles of forked. The only one of these that is phenotypically identifiable is the third. The other three can be differentiated by testing them further. The identification of these types depends on the frequencies of types recovered in a breeding test. The diagnostic crossover types in the next generation are given in figure 42. The fourth type is known as a reciprocal crossover type. As is seen in figure 42, no offspring from this type are homozygous for both the g and f alleles. Such offspring would be obtained only following a triple exchange of the proper kind. These do not occur within the regions under consideration because of interference. If the regions are long enough to give such triple exchanges, their frequency is rare as compared with singles, and there is therefore no difficulty in determining genotypes. The basis of these tests, as may be seen by study of the figures, is the determination of the combinations of alleles that become homozygous together following crossing over.

Ratio of Different Types of Double Exchanges.—We can, by taking advantage of the fact that two chromatids are recovered

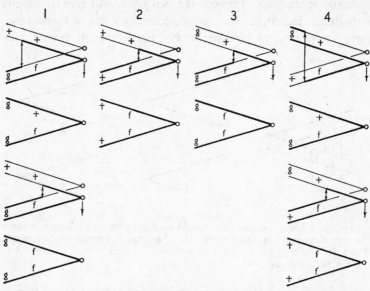

Fig. 42.—The method of distinguishing the four genotypes shown as recovered products in figure 41. In each case, only the diagnostic crossover types are shown. Recovered chromatids or chromatid segments are indicated in the tetrads (above for each pair) by heavy lines.

from a single meiotic process in attached-X females, find out whether 2-, 3-, and 4-strand doubles occur with the relative frequencies 1:2:1 expected if exchanges involve the four chromatids at random. Taking $\widehat{XX}$ females heterozygous for vermilion, garnet, and forked, as shown in figure 43, individuals can be selected from among the offspring that are homozygous for g but not for v or f. They must result from double exchanges in the two regions. There are four possible types of these as shown in figure 43. These will give two types of progeny and this can be used as a means of identifying them. Since one type is given by 4-strand doubles, the other type by 2-strand doubles, and both types in equal numbers by 3-strand doubles, an equality of the two types will indicate that 2- and 4-strand double exchanges occur with equal frequencies. The available data agree with the assumption that this is the case. By comparing the frequencies of $\dfrac{v^+ \quad g \quad f^+}{v \quad g^+ \quad f}$ with $\dfrac{v^+ \quad g \quad f^+}{v^+ \quad g^+ \quad f}$ and $\dfrac{v \quad g \quad f^+}{v \quad g^+ \quad f}$ genotypes, as determined by testing, one can compare the relative frequencies of 2- and 3-strand double exchanges. Actual data indicate a 1:2 ratio of these types.

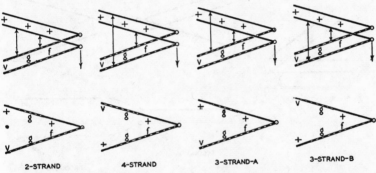

2-STRAND   4-STRAND   3-STRAND-A   3-STRAND-B

Fig. 43.—Diagrams showing methods of determining the relative frequencies of 2-strand and 4-strand double exchanges in an attached-X strain of Drosophila. The details of these are given in the text.

Attached-X stocks of Drosophila, then, have made it possible to determine several of the characteristics of the process of crossing over that we cannot find out about from a study of normal stocks. There are other examples of this same general nature—in which deviations from the normal hereditary mechanism give valuable information about the nature of the mechanism.

DETACHMENT AND THE Y CHROMOSOME

About one out of every 2,000 eggs produced by an attached-X female contains a single X chromosome. This may give rise to an exceptional female or to an exceptional male, depending on whether it is fertilized by an X- or a Y-bearing sperm. Cytological study of these exceptions shows that a single X derived from two attached-X chromosomes, a process known as *detachment*, has a part of a Y chromosome attached to it, either a short arm or a long arm. Such detachments result from rare crossing over between an X and the Y chromosome near the centromere (Fig. 44). It is assumed that near the centromere, the X and Y chromosomes are homologous, and cytological observations on the male show that these regions conjugate. This assumption is strengthened further by the fact that the sex-linked gene bobbed,

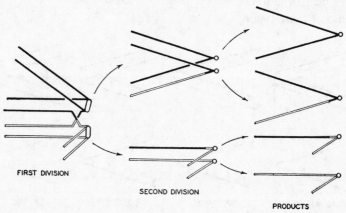

FIRST DIVISION

SECOND DIVISION

PRODUCTS

Fig. 44.—Meiotic divisions in attached-X female of Drosophila showing crossing over between one member of a pair of attached-X chromosomes and the Y chromosome.

known to be located very near the centromere, has an allele in the Y chromosome; it is the only gene known in the X chromosome of Drosophila that does have a definite allele in the Y.

Heterochromatin.—It has already been pointed out that the Y chromosome is largely empty. Such genetically empty chromatin is frequently called *inert* by geneticists and appears to be identical with the *heterochromatin* of cytologists. The argument of localized homology between X and Y chromosomes is strengthened by the fact that all of the genes in the X chromosome with the exception of bobbed lie in the distal two thirds of the chromosome as seen at somatic metaphase. In other words, the proximal one third is inert like most of the Y. Cytologically, hetero-chromatin can often be identified by the fact that under certain conditions it stains more deeply than does active or *euchromatin*. Cytologically it can be shown that there is heterochromatin near the centromeres of all the autosomes of Drosophila. Further-more, heterochromatin is widely distributed in many plants and animals.

RECOVERY OF ALL PRODUCTS OF MEIOSIS IN PLANTS

In some of the lower plants it is possible to recover all four haploid products of the two meiotic divisions of a single diploid cell, and to determine their genetic constitution individually. This

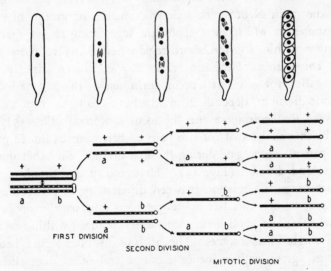

Fig. 45.—Diagram of the formation of ascospores in Neurospora. Above, the arrangement of spindles leading to a regular linear orientation of ascospores is given. Below, the behavior of one pair of chromosomes at the first and second meiotic divisions and during the subsequent mitotic division.

has been done in Sphaerocarpos, a liverwort, by Allen. An even more convenient organism for this purpose is the ascomycetous mold, Neurospora, which has been studied by Dodge and by Lindegren. In this plant, as in other Ascomycetes, the ascospores are haploid and result from meiotic divisions in a mother nucleus.

The products are arranged in a regular manner in a sac-like structure, called an *ascus,* as indicated in figure 45. The four products of meiosis go through a haploid mitotic division as in many other Ascomycetes, giving rise to eight ascospores. Thus there are always four pairs, the two members of each pair being genetically identical. The advantage of using this organism is that the ascospores can be taken out in order and cultured separately. The four at one end of the ascus were separated from the four at the other end at the first meiotic division; the two pairs of adjacent (and identical) spores within each group of four were separated at the second (Fig. 45). The spores germinate directly into haploid individuals, so that no complications result from fertilization, from dominance, etc. Several linked genes are known in Neurospora. Genetic analysis of groups of eight ascospores in which such genes are segregating shows that exchanges within a given short region never involve more than two chromatids. Furthermore, it may be shown directly that segregation for genes may occur during either the first or second meiotic divisions, depending on whether a crossover has occurred between the centromere and the locus concerned (Fig. 45). It can be shown also that for two pairs of heterozygous linked genes one pair may segregate during the first division, the other during the second division (Fig. 45). Here, too, it is a simple matter to determine the relation between different types of double exchanges, though this has not as yet been satisfactorily done, because of lack of good genetic characters localized within one arm of a single chromosome. The analysis of Neurospora heterozygotes agrees with evidence from several other organisms in showing that the process of crossing over occurs during the first of the two meiotic divisions.

REFERENCES

Beadle, G. W., and S. Emerson. 1935. Further studies of crossing over in attached-X chromosomes of Drosophila melanogaster. Genetics, 20:192–206.

Lindegren, C. C. 1936. A six-point map of the sex-chromosome of Neurospora crassa. Journ. Genet., 32:243–256.

Mather, K. 1938. Crossing over. Biol. Reviews, 13:252–292.

PROBLEMS

1. A single male showing a new character is found in Drosophila. When this male is crossed with wild-type females, the F_1 consists of wild-type individuals only. What does this tell one about the inheritance of the mutant character? If a cross between the mutant male and an attached-X females gives daughters like the mother and all of the sons like the father, what additional information does this result give one?

2. If a yellow attached-X Drosophila female is mated to a male that is white-eyed and is homozygous for the vestigial gene (vg—second chromosome), what are the genotypes and phenotypes expected among the offspring?

3. Starting with an attached-X female of the constitution $\dfrac{+\ +\ +}{y\ ct\ v}$, indicate diagrammatically the crossovers necessary to give daughters of the following genotypes:

   a. $\dfrac{y\ +\ +}{y\ ct\ v}$ c. $\dfrac{y\ ct\ v}{y\ ct\ v}$

   b. $\dfrac{y\ ct\ +}{y\ ct\ v}$ d. $\dfrac{y\ ct\ +}{+\ ct\ v}$

   e. $\dfrac{+\ ct\ +}{y\ +\ v}$

4. What would you deduce as to the genotypic constitutions (for loci $y\ ct\ v$) of an attached-X female that gave rise to the following phenotypes among her daughters (all other phenotypes with a low frequency)?

   a. $\begin{array}{c} + \\ y \\ y\ ct \\ y\ ct\ v \end{array}$ b. $\begin{array}{c} + \\ y \\ y\ ct \\ v \end{array}$ c. $\begin{array}{c} ct \\ y\ ct \\ ct\ v \end{array}$ d. $\begin{array}{c} + \\ y \\ y\ ct \end{array}$ e. $\begin{array}{c} + \\ y \\ ct \end{array}$

5. What phenotypes would you expect from attached-X females of the following genotypes?

   a. $\dfrac{y\ +\ +}{y\ rb\ cv}$ d. $\dfrac{+\ +\ +}{y\ rb\ cv}$

   b. $\dfrac{+\ rb\ +}{y\ +\ cv}$ e. $\dfrac{y\ rb\ +}{+\ +\ +}$

   c. $\dfrac{+\ rb\ cv}{y\ +\ +}$ f. $\dfrac{+\ +\ cv}{y\ +\ +}$

6. Assuming that attached-X females of the constitution $\dfrac{+\ +}{y\ \ cv}$, and wild-type males are available, outline the procedure you would use to maintain cultures from which females of this same constitution could be obtained.

7. The Y chromosome in Drosophila usually carries the normal allele of the recessive sex-linked gene bobbed. What would be the appearance of females and males in a culture in which all X's carried a recessive bobbed allele? Give the results through the F_2 generation of a cross between a bobbed female and a male from a true-breeding wild-type stock. (Note: It is possible to have a recessive bobbed allele in a Y chromosome, although this is not the usual condition.)

8. What would be the result of the reverse of a crossover that gives rise to a "detachment" of attached-X chromosomes? (See figure 44.) What method does this suggest for making up attached-X stocks of desired constitutions?

9. If a heterozygous gene a in Neurospora segregates in the second meiotic division in 10 per cent of the asci, what is the map distance between the centromere and the locus of a? If gene b shows 8 per cent crossing over with the centromere, in what per cent of the asci would it show first division segregation?

10. With the three genes a, b, and c located at 10, 20, and 25, respectively, on a chromosomes map, what would be the double crossover frequency with coincidence of 0.5? The double exchange frequency with this coincidence?

CHAPTER VIII

INTRA-CHROMOSOMAL REARRANGEMENTS: INVERSIONS

UP to this point it has been assumed that the arrangement of hereditary material is invariable within a species; that is, that the number of chromosomes is constant and that the sequence of genes within a chromosome is permanently fixed. We have, of course, considered the orderly interchange of corresponding segments of homologous chromosomes known as crossing over, but this has no effect on the arrangement of loci; it merely results in new combinations of alleles. The constancy in arrangement of chromatin is only relative; changes do occur in the sequence and grouping of loci, but such changes are rare under natural conditions. The frequency with which changes of this kind occur can be greatly increased by means of artificial radiation—x-ray treatment for example. In addition to changes in genes themselves, which will be considered in Chapter XV, x-rays result in various kinds of chromosome rearrangements or aberrations. For example, a segment of a given chromosome may be inverted with respect to its previous arrangement. If, at meiosis, one chromosome has the old arrangement and its homolog has the new arrangement, it is evident that there will be mechanical difficulties when like parts come to pair during conjugation. The study of chromosome pairing in such *structural heterozygotes,* and of the genetic consequences, gives information as to the general properties of the mechanisms concerned.

Illegitimate Crossing Over.—The exact mechanism by which chromosome rearrangements arise is not known. It has been suggested by Serebrovsky that the process by which they occur is similar to crossing over but involves non-corresponding parts of chromosomes or even non-homologous chromosomes (a process sometimes known as "illegitimate crossing over"). It is not clear just how far this analogy with crossing over can be carried, but the interpretation on this basis at least provides a convenient means of classifying chromosome aberrations. In the diagrams based on this interpretation we have indicated one difference from true crossing over, namely, that only two chromosome threads are present at the time of breakage and reunion. Presumably rearrangements may occur either before or after splitting; the chromosomes are shown as single only because the diagrams are simpler.

Types of Rearrangements.—The various possible reorganizations within a single chromosome following illegitimate crossing over are shown in figure 46. An internal loop followed by breakage and reunion along a new plane may give rise to an inversion, in which, in effect, an internal segment is removed, turned through 180 degrees, and reinserted. The inverted segment may or may not include the centromere. The same configuration, followed by breakage and fusion along a different plane, can give rise to a chromosome with a ring-segment removed. If the loop does not include the centromere, the chromosome from which the segment is removed will retain its centromere, and the closed or ring-segment will be *acentric* (have no centromere). If, on the other hand, the loop does include the centromere, the ring will retain the centromere and the rod segment will be acentric. All of these aberrations are known either genetically, cytologically or both.

The origin of aberrations involving non-homologous chromosomes is indicated in figure 46. Interchange of segments of non-homologous chromosomes is known as *reciprocal translocation* or sometimes as *segmental interchange*. Only one of the two possible planes of breakage and reunion gives rise to a translocation in which each new chromosome has a single centromere. Reunion in the other new plane gives one acentric and one *dicentric* (two centromeres) chromosome. Acentric and dicentric chromosomes or chromosome fragments are not usually transmitted from one cell generation to another with any regularity.

Inversions.—Of the various types of aberrations mentioned above and indicated in figure 46, only inversions will be considered in any detail in this chapter; the other types will be discussed later.

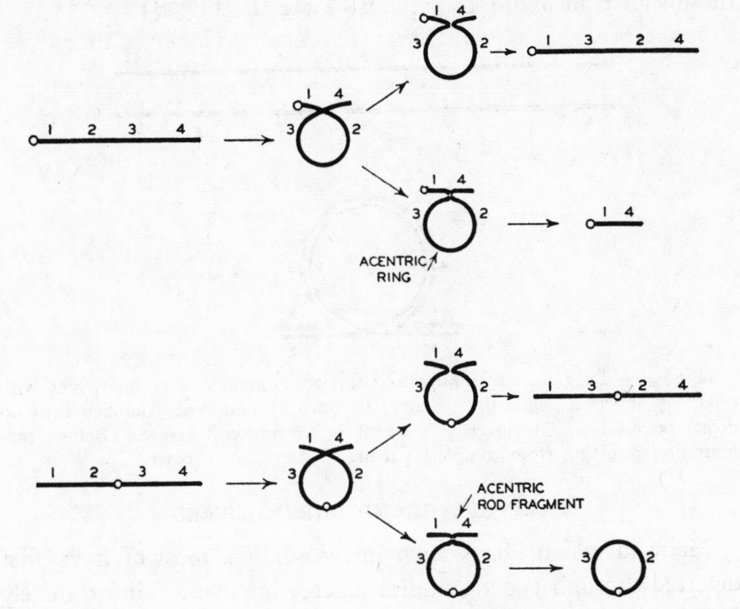

Fig. 46.—Diagram of results expected from illegitimate crossing over. From above: Production of inversion, deletion, centromere-including inversion, ring-chromosome, reciprocal translocation, and dicentric and acentric translocation chromosomes. In all cases acentric chromosomes are eliminated. (Based on Serebrovsky's scheme.)

In an individual in which one member of a pair of homologs carries an inverted segment—an *inversion heterozygote*—conjugation in the structurally heterozygous pair can be complete only if the pachytene chromosomes are thrown into loops. Cytological studies of several plants have shown that conjugation actually does occur in this way. For example, McClintock has studied inversions in maize and has found pachytene configurations such as those shown in figure 47 (see also Plate III, Fig. B).

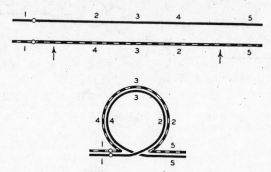

Fig. 47.—Diagram of conjugation of two homologous chromosomes with relatively inverted segments. Above, the two chromosomes numbered to indicate homology. The inverted segment is set off with arrows. Below, configuration resulting from complete pairing of homologous parts.

SALIVARY GLAND CHROMOSOMES

Many inversions are known in Drosophila, most of them first detected through genetic studies on crossing over. Unfortunately Drosophila is particularly difficult material for cytological studies of the meiotic divisions, primarily because of the small size of the chromosomes, and it has not been possible to study the process of conjugation of homologous chromosomes. It has been found, however, that the chromosomes of the salivary glands of larvae of Drosophila undergo a process of pairing very much like that characteristic of meiotic chromosome conjugation. These salivary gland chromosomes, known as a cytological "curiosity" for some fifty years and only recently fully appreciated, are extremely useful in studies of chromosome organization and of chromosome rearrangements.

Salivary gland chromosomes of the type found in Drosophila are found in most of the species of the insect order Diptera, and are unknown outside of this order. They are found in the large cells of the larval salivary glands, in which the chromosomes attain their maximum size at about the time of puparium forma-

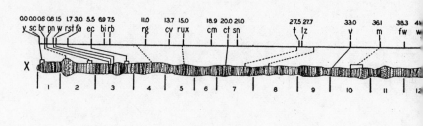

X chromosome linkage map:
0.0 0.0 0.6 0.8 1.5 1.7 3.0 5.5 6.9 7.5 11.0 13.7 15.0 18.9 20.0 21.0 27.5 27.7 33.0 36.1 38.3 4?
y sc br pn w rst fa ec bi rb rg cv rux cm ct sn t lz v m fw w

X

1 2 3 4 5 6 7 8 9 10 11 12

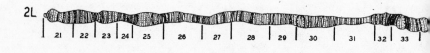

2L

21 22 23 24 25 26 27 28 29 30 31 32 33

2.R

41 42 43 44 45 46 47 48 49 50 51

3L

61 62 63 64 65 66 67 68 69 70 71

3R

81 82 83 84 85 86 87 88 89

Fig. 48.—The salivary gland chromosomes of Drosophila melanogaster. The upper line shows the linkage map of the X chromosome, with corresponding points on the salivary gland chromosome indicated. The numbered sections below each chromosome are arbitrarily chosen, for convenience in referring to

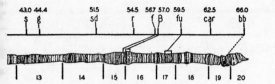

OOGONIAL
METAPHASE

4

specific regions. The chromocenter, in which the base of each chromosome is embedded, is not shown. The oogonial metaphase figure (above, to right) is drawn to the same scale. (After Bridges, Journal of Heredity.)

tion. These chromosomes are unusual in that they remain in a prophase-like condition—a characteristic that makes them particularly convenient for cytological study. The nuclei of the salivary gland cells do not take part in further mitotic divisions; the cells break down in the early stages of metamorphosis. Another important property of these chromosomes is that they are clearly differentiated by cross discs. Discs at different levels vary in size and reaction to stain, but corresponding discs in chromosomes from different cells are similar in these respects. This cross banding results in patterns that are characteristic of particular regions of particular chromosomes. These are shown in the photomicrograph reproduced in figure 49, and are indicated semi-diagrammatically in figure 48. As indicated above, the two members of a pair of salivary gland chromosomes undergo a very intimate and regionally specific pairing similar to meiotic conjugation. This pairing is so intimate that it gives the appearance of an actual fusing of the chromosomes. The strands shown in the photographs appear superficially single but they arose by pairing of two originally separate homologs. As a matter of fact, it is believed that each separate homolog is made up of many ultimate chromosome threads, and that this may explain their large size, *i.e.*, each apparent single chromosome really represents many chromosome threads lying parallel. The cross-discs may be made up of groups of homologous chromomeres.

The Chromocenter.—Examination of salivary gland chromosome preparations of Drosophila melanogaster shows what appear to be six units (Figs. 48 and 49). As indicated above these are paired. It will be recalled that somatic divisions show four pairs of chromosomes. This apparent discrepancy results from the fact that the arms of the V-shaped second and third chromosomes appear to be separate in the salivary gland nuclei. The heterochromatin near the centromeres is diffuse and more or less indistinct. Furthermore, the heterochromatin of all the chromosomes forms a single aggregate and results in the typical arrangement in which chromosome arms radiate from a common center. This aggregate of heterochromatin in salivary gland nuclei is known as the *chromocenter*. In some other Diptera, for example, Chironomus or Sciara, there is no common chromocenter, and the number of separate units in the salivary glands is the same as the number of pairs of mitotic chromosomes. The Y chromosome of Drosophila appears largely as chromocentral material in male salivary gland preparations.

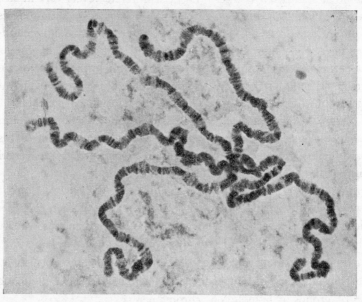

Fig. 49.—Salivary gland nucleus of Drosophila melanogaster. There are five long arms (X and right and left arms of 2 and 3) and one very short one (4), attached to a common chromocenter. Each arm is made up of two paired homologs. Progressing clockwise from the lower left of the figure, the chromosome ends are 2 right, 2 left, 3 left, X, 4 (short chromosome extending to the right of the chromocenter), and 3 right. The X is homozygous for the inversion shown in heterozygous form in Plate III, A. (Photograph from Dr. B. P. Kaufmann, to appear in *Journal of Heredity*.)

INVERSIONS

Conjugation in Inversion Heterozygotes.—The pairing of homologous salivary gland chromosomes in an inversion hetero- zygote in Drosophila is shown in Plate III, figure A. The strik- ing characteristic of pairing here is that there is an exact alignment of corresponding cross bands. Also it is clear that within the inversion loop, one homolog has an inverted sequence with respect to the other. Careful comparison with the salivary gland chromosome map shown in figure 48 shows that the inverted segment is in the X chromosomes and includes a segment extending from section 4 to section 11. It is evident that a study of the band patterns at the point where a change in homology in the two members of the chromosome pair occurs enables one to localize the inversion ends exactly. This cytological localization of inversions can be correlated with genetic behavior, and this correlation provides one method of identifying genetic linkage groups with particular chromosomes as observed cytologically. As an example,

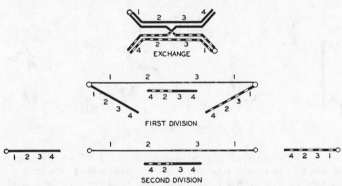

Fig. 50.—Results of single exchange between homologs with relatively inverted segments. Conjugation is indicated only for the inverted segments. Only the non-crossover chromatids give rise to whole chromosomes with one centromere.

the inversion shown in Plate III, figure A, can be shown by purely genetic means to be in the X chromosome; obviously, then, the group of sex-linked genes must be carried in the chromosome that shows the inversion configuration in the salivary glands.

Exchange in Inversion Heterozygotes.—Inversions in Drosophila were first detected by their effect on recombination frequencies; in fact, they were known simply as "crossover reducers" for many years before their true nature was understood. Crossing over between a pair of homologs heterozygous for an inverted segment should result in no serious consequences if the crossovers occur outside the limits of the inversion loop. Crossovers within the inversion loop, however, will lead to mechanical difficulties in meiosis. Indicating conjugation between inverted segments only, we see from figure 50 that a single exchange will give rise to one dicentric and one acentric chromatid. During anaphase of the first meiotic division, with the two centromeres moving toward opposite poles, the dicentric chromatid evidently forms a tie between the centromeres. Cytological studies of meiosis in inversion heterozygotes, especially in plants, show that such ties or *chromatin bridges* do occur (Fig. 51). Such studies also indicate the fate of the acentric chromatid. Since it has no centromere, it does not become attached to the spindle and consequently remains

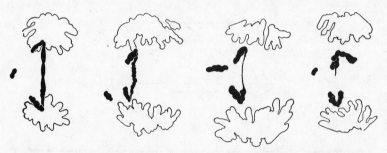

Fig. 51.—Telophase of first meiotic division in Fritillaria showing chromatin bridges and acentric fragments resulting from crossing over between segments inverted with respect to each other. The difference in the size of the fragments results from differences in positions of the different inversions involved. (After Bennett.)

behind or "lags" during anaphase separation (Fig. 51, Plate III, C, D, E, F). This failure of an acentric chromatid or chromosome to move toward one pole or the other during cell division is a general rule; acentric chromosomes formed in other ways show similar behavior.

Scute-4 Inversion.—As a more or less typical example of a long inversion in the X chromosome of Drosophila, we can take the one known as the scute-4 inversion (symbol, *In sc⁴*; the terminology for inversions has not been systematized—this name comes from the fact that this inversion is associated with a scute allele known as scute-4). This inversion involves the greater part of the active material of the X chromosome; it extends from

PLATE III

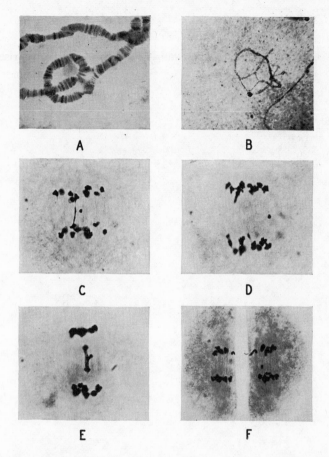

A

B

C

D

E

F

(All photographs of maize chromosomes by kindness of Dr. Barbara McClintock.)

A, Salivary gland chromosome configuration of delta-49 inversion heterozygote in the X-chromosome of D. melanogaster. (Photograph by Miss Margaret Hoover.)

B, Pachytene configuration in a maize plant heterozygous for an inversion in chromosome 2. Note the loop configuration shown here.

C, Anaphase bridge and acentric fragment resulting from crossing over between relatively inverted segments of chromosome 4 of maize.

D, A later stage than shown in C. The bridge has broken, but the acentric fragment remains between the two groups of chromosomes.

E, A double bridge and two acentric fragments (not lying free of the bridges) following a 4-strand crossover between the same relatively inverted segments as are concerned in C and D.

F, Second meiotic division anaphase in chromosome 4 inversion heterozygote. The unequal products of breakage of the bridge are seen near the inner cell boundaries in the two cells. The acentric fragment lies in the right cell just above the center.

just to the right of the scute locus to between carnation and bobbed (see chromosome map, Fig. 32, and salivary gland chromosome chart, Fig. 48). The progeny of a female heterozygous for this inversion show no single crossovers. Determinations of egg and larval mortality fail to show a frequency of inviable zygotes sufficient to account for the single crossovers expected to occur.

Elimination of Single Crossover Chromatids.—There are two possible explanations of this apparent discrepancy, either: (1) single exchanges do not occur with any appreciable frequency or

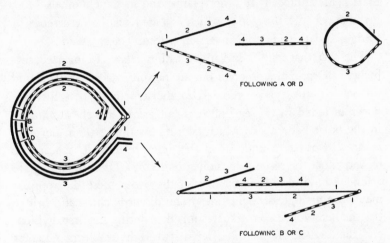

FOLLOWING A OR D

FOLLOWING B OR C

Fig. 52.—Results of single exchange between the two members of a pair of attached-X chromosomes that differ by a long inversion. From exchange A or D either a ring-chromosome or a non-crossover pair of chromosomes is recovered. From exchange B or C a chromatid tie results which, since it occurs at the second meiotic division, presumably results in an inviable egg.

(2) they, for some reason, are not included in the egg. Breakage of the dicentric chromatid would be expected to result in a broken chromosome which would either result in the death of the zygote or be detected genetically. The first point can be tested by deliberately arranging conditions in such a way that, if single exchanges do occur, some of the products will have a single centromere. This can be done by using attached-X females heterozygous for an X chromosome inversion. Here there is a single centromere for two chromosomes, and a single exchange within

the inversion can give rise to a ring chromosome with an extra segment near the centromere and a piece missing at the end (Fig. 52). If the extra piece and the missing piece are small, the ring, in combination with a normal X chromosome, will give rise to a viable female. The frequency of these "exceptions" to the attached-X type of sex-linkage shows that single crossing over within a long inversion is not much different from that characteristic for the same segment of chromosome normally arranged. This indirect genetic evidence agrees with the cytological evidence in maize (McClintock), in other plants, and in some animals (*e.g.*, grasshoppers, Darlington), which is based on the occurrence of bridges and fragments in inversion heterozygotes.

The fact that single exchanges within relatively inverted segments in Drosophila do occur but are not ordinarily recovered and do not kill the zygote, has led to the hypothesis that they are never included in the egg nucleus but are selectively eliminated in the polar body nuclei which take no essential part in development. In order to understand this interpretation it is necessary to know that the second division spindles in a Drosophila egg are arranged in tandem, *i.e.*, the axes of the two spindles lie approximately on a single straight line. The dicentric chromatid bridge ties the homologous centromeres together during the first division, and the acentric chromatid is left between the resulting daughter nuclei (Fig. 50). At the second division, the non-crossover chromatids are free to move into the terminal nuclei (Fig. 50). Since it is always one of these terminal nuclei that functions as the egg nucleus (the one farthest from the surface of the egg), it is evident that only a non-crossover chromatid can be recovered.

Inversions and Patroclinous Males.—Considering now the three possible types of double exchanges within an inversion, it is evident that a 2-strand double results in no difficulty. Disjunction is normal, and there is an equal chance of recovering a non-crossover or a double crossover chromatid in the egg nucleus (Fig. 53). On the other hand, a 3-strand double gives the same geometrical configuration as does a single exchange (Fig. 53). Of the three types of double exchanges, the 4-strand double is particularly in-

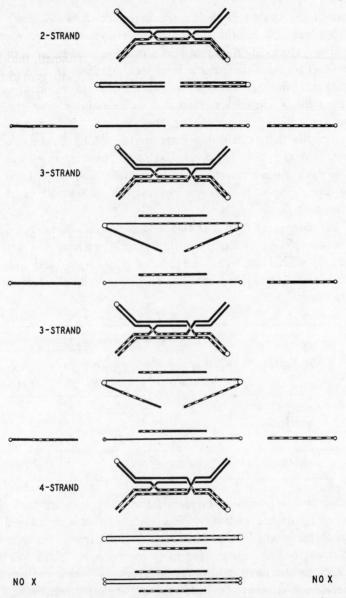

Fig. 53.—Diagrams showing the consequences of double exchanges between relatively inverted segments in X chromosomes of Drosophila. Each type of double is diagrammed in three successive stages, reading downward. Either the right or left product of the second divisions may be included in the nucleus that functions as the egg nucleus.

teresting. As seen in figure 53, it gives rise to two dicentric and two acentric chromatids. Such configurations have been seen in plants. Following the assumptions made in connection with single exchanges, all four chromatids should remain in the two central nuclei during the second meiotic division; that is, nothing, so far as the X chromosome is concerned, should get into the egg nucleus. This is exactly what does happen. It can be shown genetically that females heterozygous for a long X chromosome inversion give rise to *patroclinous* males, that is, males that received an X chromosome from their father and no X from the mother. The eggs that give rise to such exceptional males are known as "no-X" eggs.

Inversions and Types of Double Exchanges.—If we put down the expected products following double exchanges involving strands at random (*i.e.*, a 1:2:1 ratio of 2-, 3- and 4-strand double exchanges), we have the following, with the frequencies indicated:

Exchange	Recovered in egg nucleus		
	Non-crossover	Double crossover	No-X egg
2-strand	1	1	
3-strand	2	2	
4-strand			2
Total	3	3	2

Since double crossover chromatids and no-X eggs are both produced by double exchanges, the relation between them will be fixed (there will be a small correction for triple exchanges, but this will be insignificant for most inversions). Since half the no-X eggs are fertilized by sperms carrying a Y chromosome and therefore give rise to inviable eggs, the indicated relation will be expressed, in terms of zygotes, as a ratio of three recovered double crossovers to one patroclinous male. This ratio agrees with the experimental results for X chromosome inversions. A further check on this general interpretation is the fact that egg-inviability corresponds approximately in frequency to patroclinous males.

The behavior of inversion heterozygotes in Drosophila provides, in a rather indirect way, confirmation of the assumption that the crossover strands in multiple exchanges are involved at random; that is, that the strands which cross over at one level do not prejudice those taking part in another exchange.

Inversions that Include the Centromere.—Crossing over within an inversion loop in which the centromere is included in the inversion will not give rise to acentric and dicentric chromatids, but results in complications of another kind as shown in figure 54. Here single exchanges give rise to two aberrant chromatids. One

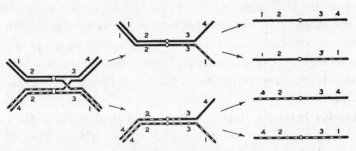

Fig. 54.—Complementary deficiency-duplications resulting from a single exchange between relatively inverted segments including the centromere. Conjugation of terminal segments is not indicated.

of these will have one section of chromosome represented twice (a *duplication*) and another section not represented at all (a *deficiency*). The other aberrant chromatid will be the complement of the first. Two chromatids are normal, one with the original sequence of loci and one with the inversion sequence. The fate of the duplication-deficiency chromosomes that result from such exchange will depend on both the length of the inversion and the organism concerned. In maize, the only organism in which such an inversion has been studied both genetically and cytologically, these chromatids result in inviability of the gametophyte (Müntzing and Anderson).

Inversion Homozygotes.—It is possible to replace the alleles originally present in an inversion by taking advantage of the fact that double crossovers can be recovered. In this way individuals homozygous for an inversion but heterozygous for genes within the inversion can be studied. A determination of the sequence of loci in a stock homozygous for an inversion shows that this sequence has been changed in relation to the original sequence. Thus, if the original sequence is 1-2-3-4-5, the new arrangement may be found to be 1-4-3-2-5. This is, of course, a convincing demonstration that genes are arranged in a linear order in the chromosome. All loci within the inversion have their sequence changed in relation to those outside the inversion. From this it is clear that in the example given, the inversion ends lie between 1 and 2 at one end and between 4 and 5 at the other. Comparisons of this kind between the position of inversion ends in terms of the genetic map and their positions in the salivary gland chromosomes, obviously can be used to localize genes in the physical chromosomes. Other methods of doing this are more efficient than this one, however; some of them will be discussed in later chapters.

Double Inversion Heterozygotes.—It is possible for a pair of homologs to differ by two or more inversions. Thus in Drosophila pseudoobscura, Dobzhansky and Sturtevant report differences of as many as five inversions between two third chromosomes. The cytological analysis of cases as complex as this is of course somewhat difficult. The genetic consequences of more than a single inversion difference between two homologs will depend on their relation to one another, whether *independent* (involving separate segments), *overlapping* (each having a segment common to the other in the inversion, as well as a segment that is not common), or *included* (one inversion entirely within another).

Detecting Inversions.—From the foregoing account it is evident that any one of a number of means can be used to detect inversions. Direct cytological detection of the characteristic inversion loop at pachytene or in the salivary gland nuclei has been used, and this method has one important advantage—very short inversions, difficult to detect in other ways, can be seen. This method is extensively used in studying Drosophila species. Another cytological method of detecting inversion heterozygotes involves the observation of chromatin bridges and accompanying fragments at the anaphase of the first meiotic division. Darlington and others have shown, by the use of this method, that the frequency of inversion heterozygotes may be relatively high in wild populations. Inversions can be detected by determining the order of genes in the homozygote. This method is laborious and time-consuming, is not often used, although it was the original method of proving the existence of inversions. The apparent reduction in crossing over in inversion heterozygotes has been used frequently in detecting such rearrangement of chromatin. This is the most useful genetic method, but because crossing over can be reduced in other ways, it requires confirmatory evidence.

REFERENCES

Darlington, C. D. 1937. Recent advances in cytology. 671 pp. Blakiston's Sons and Co., Philadelphia.

Dobzhansky, T., and A. H. Sturtevant. 1938. Variations in the gene arrangement in the chromosomes of Drosophila pseudoobscura. Genetics, 23:28–64.

Sturtevant, A. H., and G. W. Beadle. 1936. The relation of inversions in the X chromosome of Drosophila melanogaster to crossing over and disjunction. Genetics, 21:554–604.

PROBLEMS

1. Draw the configuration expected with complete conjugation of all parts of two homologs differing by one inverted segment, *e.g.*, 1 2 3 4 5 6 7 8 and 1 2 · 6 5 4 3 · 7 8, where centered periods indicate breakage points.

2. What types are expected in the next generation following self-pollination of an inversion heterozygote in maize, *e.g.*, $\dfrac{1\ 2\ 3\ 4}{1\ 3\ 2\ 4}$? Give relative frequencies of the various types.

3. The inversion in Drosophila known as *In scute*-7 lies at the left end of the X chromosome. Females heterozygous for this inversion show no recovered crossovers for yellow and cut. Singles between cut and singed are recovered rarely, while crossing over to the right of singed is normal. What per cent of crossing over would you expect between yellow and vermilion in an *In scute*-7 heterozygote? (Look up loci on chromosome map.)

4. The five genes *a*, *b*, *c*, *d*, and *e* are linked in the order given and give the following crossover values:

$$a - b \quad 10 \qquad c - d \quad 5$$
$$b - c \quad 5 \qquad d - e \quad 10$$

Assuming that the frequency of crossing over for each of these intervals remains unchanged, draw a chromosome map for an inversion homozygote in which the inversion breaks occurred just to the left of *b* and just to the right of *d*.

5. Starting with an inversion in a wild-type stock, what method could you suggest for getting mutant genes into the inverted segment of the inversion-carrying chromosome? (For example, starting with *a b c d e f*, and the inversion *a+ e+ d+ c+ b+ f+*, show how to get the chromosome *a+ e+ d c b+ f+*.)

6. Draw the configuration expected following complete conjugation of all parts of two homologous chromosomes that differ by two independent inversions, *e.g.*, 1 2 · 5 4 3 · 6 7 8 9 10 11 12 and 1 2 3 4 5 6 7 · 10 9 8 · 11 12. (It is suggested that heavy cord, modelling clay, or insulated copper wire can be used to advantage for making models of configurations of this kind.)

7. Draw the configuration expected with complete conjugation between two homologous chromosomes that differ by two inversions, one entirely included within the other, *e.g.*, 1 2 · 10 9 8 7 6 5 4 3 · 11 12 and 1 2 3 4 · 8 7 6 5 · 9 10 11 12.

8. Draw the configuration expected with complete conjugation between two homologous chromosomes that differ by two overlapping inversions, *e.g.*, 1 2 · 8 7 6 5 4 3 · 9 10 11 12 and 1 2 3 4 · 10 9 8 7 6 5 · 11 12.

9. A normal chromosome map with four loci is *a* 10 *b* 25 *c* 20 *d*.

A four-point testcross, $\dfrac{+\ +\ +\ +}{a\ b\ c\ d} \times a\ b\ c\ d$ gives the following data:

+	+	+	+	360	+	b	+	+	0
a	b	c	d	350	a	+	c	d	0
+	b	c	d	42	+	b	c	+	4
a	+	+	+	48	a	+	+	d	6
+	+	c	d	0	+	+	c	+	0
a	b	+	+	0	a	b	+	d	0
+	+	+	d	92	+	b	+	d	0
a	b	c	+	98	a	+	c	+	0

How would you interpret these data on the assumption that an inversion is involved?

10. If you desired to get males without a Y chromosome in Drosophila, *i.e.*, patroclinus XO males, by use of an inversion, what type of an inversion would you select for the purpose? Give reasons.

11. The following three sequences are found for a single chromosome in different strains:

(1)   1 2 3 4 5 6 7 8
(2)   1 2 5 4 3 6 7 8
(3)   1 2 3 6 5 4 7 8

Assuming that sequence (2) was the original one, what can be concluded concerning the order in which the other two arose, and concerning their relationship to each other?

CHAPTER IX

INTRA-CHROMOSOMAL REARRANGEMENTS: INCOMPLETE CHROMOSOMES

DEFICIENCIES

A *deficiency* may be defined as the absence of any chromosome section from a haploid set. There are various ways of obtaining aberrations of this type, as we shall see. Illegitimate crossing over, as explained in Chapter VIII, may give rise to a rod chromosome and an acentric ring. The rod chromosome has had an internal section removed (the ring) and the ends joined. A de-

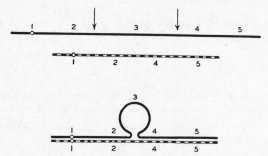

Fig. 55.—Scheme of conjugation between two homologs, one of which is deficient for a segment. The extent of the deficiency is indicated by arrows in the upper unpaired chromosome.

ficiency of this particular type in which an internal segment is absent is called a *deletion*. The conditions responsible for the origin of a deletion may be thought of as being similar to those responsible for the origin of an inversion with the difference that, instead of the segment being inverted, it is left out.

Conjugation in Deficiency Heterozygotes.—It is clear that two homologs that are heterozygous for a deletion (*i.e.*, one normal chromosome and one from which a section has been deleted) will have, corresponding to the missing section, a segment that has nothing with which to conjugate. If the parts that do correspond conjugate completely, the segment of the normal homolog for which there is no pairing partner will form an unpaired loop as shown in figure 55. Such configurations have been studied in the pachytene stage of meiosis in maize. Deletions vary in size, from those involving very small segments to those in which an appreciable part of a chromosome arm is gone. Measurements, at pachytene, of the length of the unpaired loop in a deletion heterozygote tell one exactly how large a piece was deleted. Similar cytological studies of deletions have been made in Drosophila, in the salivary gland nuclei. Because of the distinctness of the cross-bands in these chromosomes it is possible to detect extremely short deletions, even those in which only a single band is missing.

Terminal Deficiencies.—Deficiencies of terminal segments of chromosomes have been reported in maize, in Drosophila, and in other forms. In Drosophila they are relatively infrequent as compared with internal deletions. In fact, they are so rare that it may be questioned whether true terminal deficiencies really exist at all. There are certainly instances in which careful cytological study indicates the loss of a terminal section. The difficulty here is that it is almost impossible to be sure that there is not a very small end segment still present. If deficiencies arise by illegitimate crossing over, it is somewhat difficult to imagine how a terminal deficiency could be produced. On the other hand, simple breakage of a chromosome with loss of the acentric segment would give rise to such deficiencies. Several deficiencies that give every appearance of being terminal are known in maize.

Phenotypic Effects of Deficiencies.—The genetic characteristics of deficiencies are interesting and useful in several respects. Many of them are viable in the heterozygous condition, and they may result in marked phenotypic modifications. In Drosophila only relatively short deficiencies are viable in the heterozygous form. For example, a deficiency for half the X chromosome, even in the presence of an entire normal homolog, results in death of the individual at an early stage of development. The length of a segment that can be absent in one haploid set of the zygote and still allow it to survive depends on the location of the segment. Apparently any part or all of the Y chromosome in a male can be deficient without serious consequences in somatic development. Often the phenotypic effects of a heterozygous deficiency are quite specific, for example, the absence of a short segment, including the locus of facet, in one of the X chromosomes of a female, gives rise to an adult with a characteristic phenotype called "notch." Deficiencies for certain other short sections, when heterozygous, result in a characteristic shortening of the bristles of the fly and are known as "minutes." Often the viability of a deficiency heterozygote is much reduced, particularly in those involving a relatively long segment.

Absence of Genes in Deficient Chromosomes.—Granting that genes are strictly localized, it should follow that the removal of a definite segment of a chromosome would remove the genes within that segment. Thus, starting with a wild-type X chromosome, the removal of sections near the left end should remove the wild-type alleles of one or more genes located in this region —unless, of course, the segment removed happens, by chance, not to include any known locus. Actually it can be demonstrated experimentally that genes can be removed by removing sections of the physical chromosome. In an x-ray experiment, H. Slizynska induced a deficiency involving fifteen bands in section 3 at the left end of the X chromosome of Drosophila. This deletion-carrying chromosome, known to have carried the normal allele of the white gene prior to the treatment, was found to have no normal allele of white. From this, it is possible to say that the white locus is somewhere within this fifteen-band segment. By studying a large number of such deletions, it is possible to localize genes more exactly than this. The results of studies of a series of deletions in the left end of the X chromosome are shown diagrammatically in figure 56. In this series, every deletion in which band 3-c-1 was missing gave genetic results indicating that the normal allele of white was absent. Since no other band was deficient in all of these deletions, it follows that the white locus must be in or near this particular band of the chromosome. In the same way other genes can be localized, as a study of figure 56 will show.

Localization of Genes.—Similar methods can be used in other organisms for localizing genes within a specific chromosome and in particular regions of the chromosomes. McClintock was able

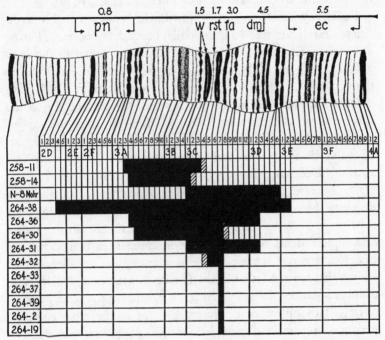

Fig. 56.—Chart summarizing the cytological and genetical characteristics of 13 deficiencies in the prune-echinus region of the X chromosome of Drosophila melanogaster. The genetic map is shown above. On the diagram of the salivary gland chromosome segment, bands are designated, for reference purposes, by a system of numbers and letters. The various deficiencies are given arbitrary designations in the column to the left. Black areas indicate regions known to be deficient, shaded areas those for which definite evidence could not be obtained. (From Slizynska.)

to show, by studying deficiencies at the left end of chromosome 2 of maize, that the liguleless (*lg*) locus must be very near the end of the short arm of this chromosome—within the last four chromomeres, in fact. In a similar way it was shown that the japonica (*j*) gene in maize is localized near the end of the long arm of chromosome 8. Other examples of a similar nature might be given.

The fact that the above method of localizing genes gives consistent results in a great many independent instances, is of especial interest in that it provides strong confirmation of the general theory of the strict localization of discrete hereditary units.

Deficiency Homozygotes.—Individuals homozygous for a deficiency rarely survive. In Drosophila, many deficiencies are known and only a few have been found that are not lethal to the individual. In general, zygotes homozygous for deficiencies die in the egg stage. There are, however, several cases in which individuals completely deficient for a small segment of the X do survive in occasional males. One of these involves the loss of the yellow locus. It should be emphasized that all of these deficiencies are very short, there being only one or two known loci absent. Deficiency homozygotes are seldom obtained in seed plants because of inviability of the gametophytes containing a deficient chromosome. This inviability of haploid gametophytes deficient in a chromosome segment is comparable to the inviability of a diploid deficiency homozygote in animals; in both cases certain loci are not represented. In maize there are deficiencies known that are not lethal in the female gametophyte, but in most of these instances, the same deficiency is lethal in the male gametophyte. This difference is presumably due to the more complex history of the male gametophyte which, of course, includes the production of a pollen tube. Another factor of importance in the transmission of deficiencies through the pollen of plants is the matter of competition in pollen tube growth. Pollen grains carrying a deficient chromosome, though potentially functional, may produce tubes with a slow growth rate and therefore never function in competition with normal pollen grains.

The fact that practically all deficiencies are lethal in haploid gametophytes of plants or in homozygous condition in the diploid is of particular interest since it leads to the conclusion that, with few exceptions, there must be at least one representative of each locus present for development to proceed normally. In Drosophila, those exceptional cases in which an individual entirely deficient for a short section of chromosome may survive, show that there is usually a very marked effect on viability as well as on particular developmental reactions. We shall return to this general question in other connections later.

DUPLICATIONS

A chromosome from which a piece has been deleted may, instead of replacing a normal chromosome of one set, be present as an extra chromosome fragment. Any extra chromosome fragment, regardless of how it originated, is known as a *duplication*. A duplication may be independent of other chromosomes, that is, have its own centromere, or it may be attached to another chromosome.

Phenotypic Effects of Duplications.—Duplications in Drosophila have phenotypic effects more or less in proportion to their lengths. Short ones may have very slight effects. Longer ones have progressively stronger effects—usually a roughening of the eyes, change in shape of the wings, modifications of bristles, and so forth. The effects depend also on the material present in the duplication, that is, in the particular segment involved. For the X chromosome and for the two large autosomes, duplications involving more than a small portion of the chromosome are usually lethal, even if present in only a single dose. Thus, a chromosome from which a segment was removed can be present in a fly as a deletion only if the deficient section is relatively short, and as a duplication only if the deficient segment includes the greater part of the chromosome.

Methods of Detecting Duplications.—In practice, duplications in Drosophila have often been obtained by mating *x*-rayed males to attached-X females homozygous for several sex-linked recessives. An $\widehat{XX}$ egg fertilized by a sperm carrying an X chromosome fragment (a deficiency in the X chromosome of the male) may give rise to a duplication-carrying female. This will be detected genetically if the duplication carries normal alleles of the genes recessive in the female. Thus, if the attached-X female is yellow and the duplication carries the wild-type allele of y, the resulting duplication female will be not-yellow. This represents a general genetic property of duplications. Any wild-type alleles present in the duplication may be expected to "suppress" the effects of any recessives present in regions of the normal chromosome corresponding to the duplication.

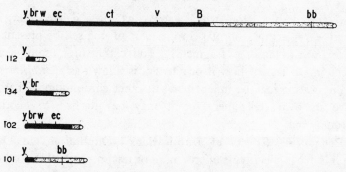

Fig. 57.—Diagrammatic representation of four X chromosome duplications in Drosophila melanogaster. An entire chromosome is shown above, active regions shown in black and inert regions (heterochromatin) in stippling. Centromeres indicated by open circles. (Based on figures by Dobzhansky.)

Figure 57 represents a number of X chromosome duplications obtained in Drosophila melanogaster by Dobzhansky. These were obtained as a result of x-ray treatment, and all of them are clearly chromosomes from which large sections have been deleted.

INVERSION CROSSOVERS

There are other ways of obtaining both duplications and deficiencies. As was pointed out in the previous chapter, crossing over in an inversion heterozygote in which the inverted segment includes the centromere, will give a combination of a duplication and a deficiency, *i.e.*, a duplication for one end of the chromosome and a deficiency for the other end. Crossing over between homologs differing by two inversions that have a common inverted region, may give rise to either duplications or deficiencies, or both. Two independent inversions, that is, those involving different segments, are of no interest in this particular connection. The results of crossing over within the segments common to the two different inversions are shown diagrammatically in figure 58. Two inversions, one included within the other, can give rise to two complementary duplication-deficiencies, each deficient for one segment and carrying a duplication for another. Overlapping inversions can give rise to double duplications or double deficiencies. Two inversions with one breakage point in common can give either a single duplication or a corresponding single deficiency. All of these types have been obtained in Drosophila. Like other duplications and deficiencies, they may be inviable in heterozy-

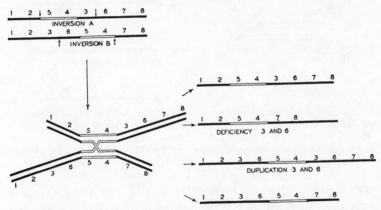

Fig. 58.—Consequences of an exchange, with the common inverted segment, between homologous chromosomes that differ by two overlapping inversions. The extent of the two separate inversions is indicated by arrows in the unpaired chromosomes at the upper left. The segments common to the two inversions are represented in outline.

gous condition, depending on the extent of the deviation from the normal amount of chromatin and also on the particular regions involved. These types, especially those in which small segments are deficient, are very useful in determining the location of inversion ends in terms of the genetic chromosome map. For example, the two inversions in the X chromosome of Drosophila known as *In scute-4* and *In scute-8* give two types of crossover products, one of them a deficiency for bobbed and no other known locus and a duplication for the scute locus. The complementary type is a duplication for bobbed and a deficiency for scute. As can be seen from figure 59, this tells us that the left break in *In sc-4* is between scute (0.0) and silver (0.1), and the right one between carnation (62.5) and bobbed (66.0); that the left breaking in *In sc-8* is between yellow (0.0) and scute (0.0) and the right one between bobbed (66.0) and the centromere

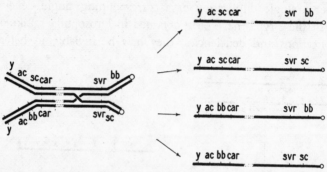

Fig. 59.—Schematic representation of the results of crossing over between the two X chromosome inversions, *In scute-4* and *In scute-8,* in Drosophila. A large proportion of the chromosomes, common to the two inversions, is omitted in the diagram; this is indicated by the dotted outlines. The crossover products are complementary duplication-deficiencies involving the scute and bobbed loci.

(exact map position not known, but very close to bobbed). The same general technique has been made use of in locating the ends of other inversions in Drosophila.

RING CHROMOSOMES

Ring-X in Drosophila.—Ring chromosome fragments such as are indicated as arising by illegitimate crossing over (Fig. 46) are known in Drosophila, in maize and in several other plants. Their properties are of importance in connection with several problems of considerable theoretical interest.

The ring-X chromosomes of Drosophila (usually referred to in the literature as "closed chromosomes" and designated by the symbol X^c) are so nearly complete chromosomes that for our purposes here they will be so considered. The two known ring-X chromosomes (X^{c-1} and X^{c-2}) each have a very short deficiency, in one case, at least, clearly to the left of the leftmost known locus (yellow). There is likewise a short duplication near the chromocenter (Schultz and Catcheside). In both cases the ring chromosomes arose as "detachments" in attached-X females. This can be pictured as illegitimate crossing over between the very tip of one homolog and the base of the other. Ring-X males and homozygous ring-X females are viable. This then represents a case of a very short non-lethal deficiency involving material which contains no known gene locus.

Crossing Over in Ring-X Heterozygotes.—Except for complications due to crossing over at meiosis, ring-X chromosomes in Drosophila behave essentially like normal ones. In a ring chromosome heterozygote a single exchange leads to a dicentric chromatid which forms a tie at the first division. This results in the elimination of the crossover products in the polar bodies in essentially the same way as does a tie following single exchange in an inversion heterozygote. Single exchange tetrads therefore result in the production of eggs with either a non-crossover rod-X or a non-crossover ring-X. Two-strand doubles result in no irregularities. Four-strand double exchanges give no-X eggs. The results of three-strand exchanges are somewhat complex and will not be considered in detail. One type of three-strand double leads to the formation of a second division chromatid tie. If this involves a centromere in the egg nucleus, it apparently results in the death of the egg.

Crossing Over in Ring-X Homozygotes.—Single exchange in a ring-X homozygote gives rise to a dicentric ring chromatid of double size which forms a tie in the first meiotic division. Again only non-crossover chromatids are free to move into the terminal nuclei, and consequently only these are recovered in the egg nuclei (Fig. 60). Multiple exchanges lead to more complex results.

Ring-fragments in Maize.—A number of informative cases of ring-fragments have been studied in plants, especially by McClintock in maize. A ring-fragment may be present in addition to a normal diploid complement of chromosomes. In this event, it behaves in many respects like other duplications. It is possible, for example, to have a plant in which the two normal fifth chromosomes carry the recessive gene brown midrib-1 (bm_1) and a ring-fragment carrying the normal allele of this gene. The ring duplication "suppresses" the brown midrib character (brown pigment in lignified cell walls). Plants of the genetic constitution just described are not uniformly wild-type (bm^+) in pigmentation but have definite sectors, usually many, that show the brown

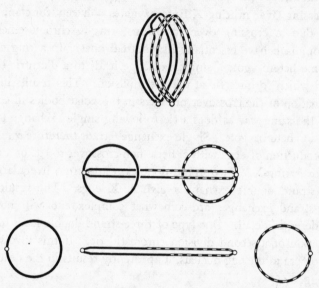

Fig. 60.—Consequences of single exchange between two ring-X chromosomes in Drosophila. Successive stages of meiosis indicated by reading downward. The dicentric double ring is presumably eliminated in the polar bodies of the egg.

midrib character. Cytological examination shows that this variegation for the brown midrib character is the result of the occasional loss, through irregular distribution, of the ring-fragment during somatic mitoses. All tissues descending from a cell in which the ring-fragment was lost have no normal allele of the bm_1 gene and therefore form a sector that shows the recessive character. This same type of variegation has been found for other characters in maize.

Size Changes in Ring-fragments.—Another characteristic of the ring-fragments studied by McClintock is that they occasionally change in size during development of the individual. The mechanism by which this occurs involves the formation of dicentric rings of double size and subsequent breakage and reunion of these at anaphase or telophase (Fig. 61). The formation of double

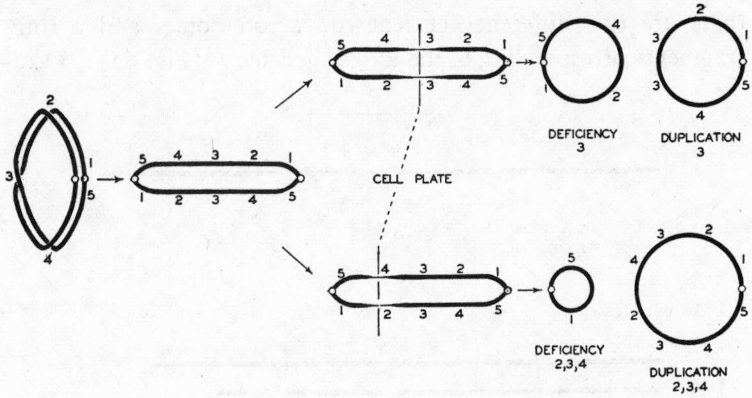

Fig. 61.—Increase and decrease in size of a ring-chromosome fragment. As indicated, the amount of change is determined by the position of the dicentric double ring-fragment relative to the cell plate. (Based on figures by McClintock.)

sized dicentric rings is not clearly understood; it may be due to irregularities in the plane of division of the ring thread or may involve a process analogous to crossing over. The end results are the same with either hypothesis. The double size dicentric ring is stretched between the division poles and appears to be cut through by the cell plate formed during telophase. The broken ends apparently always rejoin, since rod- or U-shaped fragments are never formed from ring-fragments. The breaking of the dicentric ring-chromosome may occur in such a way that the two rings formed are of unequal size (Fig. 61). This, of course, means that the original ring-fragment may be either increased or decreased in size.

Phenotypic Effects of Deficiencies.—The change in size of a ring-fragment has been used to advantage by McClintock in studying the phenotypic effects of small deficiencies. In two cases involving chromosome 5 a ring-fragment and its complement, a deleted rod-chromosome, were both recovered. This recovery of both products of the breaks was possible because of the fact that one breakage point was within a centromere, as shown in figure 62. Each of the two ring-fragments obtained carries the normal allele of the brown midrib gene. Plants can be obtained in which there are two different deficient rod chromosomes and a ring-fragment corresponding to the longer deficiency (Fig. 62). Occa-

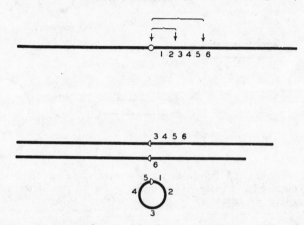

Fig. 62.—Positions (above) of breaks in chromosome 5 of maize that gave rise to rod deletions and ring-fragments studied by McClintock. As indicated by brackets, two separate alterations are involved. Below is shown the constitution of plants used in studies described in the text. Two different deletion rods and a ring-fragment complementary to one of these are present. The second ring-fragment, complementary to the other rod, was obtained but not used in the experiments considered here. (Based on figures by McClintock.)

sional decreases in the size of the ring-fragment give sectors in which there is a net deficiency. Some of these sectors are phenotypically brown midrib, presumably due to the loss of the $bm+$ allele from the ring-fragment. Other sectors have different phenotypic appearances (*e.g.*, one with colorless cell walls and white plastids), and these can be interpreted on the assumption that specific small sections of the ring-fragment are lost and that these have specific properties. Such specific properties are presumably due to specific genes in each of the sections (possibly a section corresponds to a chromomere). It is possible to deduce from the behavior of these deficient sectors something concerning the role in development of genes normally present in these segments. Thus we can conclude that at least one gene necessary for the normal development of chlorophyll must be deficient in sectors in which the plastids are colorless. The fact that sectors totally deficient for the brown midrib locus are phenotypically similar to corresponding tissues in plants homozygous for the recessive allele of the brown midrib gene strengthens this argument. Somewhat comparable situations are known in Drosophila. Males deficient in the yellow locus have been obtained and are phenotypically yellow like males with the recessive y allele present at this locus. Similarly males deficient for the scute locus, although they rarely survive, are extreme scute phenotypically. These examples raise a number of significant questions concerning the nature of genes and the difference between normal and mutant alleles; we shall return to these in later chapters.

REFERENCES

Dobzhansky, T. 1936. Induced chromosomal aberrations in animals. pp. 1167–1208, in Duggar, B. M., et al., Biological effects of radiation. McGraw-Hill Co., New York.

McClintock, B. 1938. The production of homozygous deficient tissues with mutant characteristics by means of the aberrant mitotic behavior of ring-shaped chromosomes. Genetics, 23:315–376.

Muller, H. J. 1932. Further studies on the nature and causes of gene mutations. Proc. 6th Internat. Congr. Genetics, 1:213–255.

PROBLEMS

1. Draw the configuration expected following conjugation between two homologs, one of which is deficient for a segment, *e.g.*, 1 2 3 4 5 6 and 1 2 · 5 6.

2. As indicated in the text, a specific sectional deficiency in one X chromosome of a female Drosophila results in a notch wing. Notch deficiencies do not survive in males. What sex ratio should a female heterozygous for a notch deficiency give when mated to a wild-type male?

3. From a wild-type X chromosome in Drosophila a deficient chromo-

some is obtained. When this is present with a scute white chromosome in a female the fly is phenotypically scute and white. What would you expect such a *Def/sc w* female to give in the next generation when mated to a prune (*pn*) male? When mated to a miniature (*m*) male? (Look up on the chromosome map the locations of the genes concerned. Assume that males with deficient X chromosomes do not survive.)

4. If an extreme allele of the third chromosome gene bithorax (*bx*), lethal when homozygous, appeared in a stock of Drosophila and you suspected this to be a deficiency, what genetic test could you suggest that might confirm this suspicion? (See chromosome map for positions of bithorax and of other loci that might be used in such a test.)

5. If a small independent fragment (with its own centromere), carrying the normal allele of yellow body were present in a stock of flies and were transmitted to half the gametes of individuals carrying it (assume normal distribution of whole chromosomes), what would be the phenotypes expected (including sex), and their frequencies, from the cross of duplication y/y ♀ $\times$ y ♂ ?

6. With the duplication given in the preceding problem, what would be the result of the cross of duplication $\dfrac{y \quad +}{y \quad w}$ ♀ $\times$ $y\ w$ ♂ (*i.e.*, give phenotypes and frequencies)?

7. A deleted X chromosome in Drosophila carries the normal alleles of yellow, scute, and bobbed, and can be present in either females or males. One possible method of keeping this chromosome is as a duplication in the stock yy (attached-X) duplication ♀ $\times$ y ♂ . Not-yellow females and yellow males are selected in each generation. In duplication-carrying females of this kind, crossing over occasionally occurs between the duplication and one of the attached-X chromosomes in the bobbed region. What would be the effect of such crossing over on this scheme for keeping the duplication?

8. How would you distinguish genetically between a small free duplication of unknown origin in Drosophila which resulted in a roughening of the eyes, and a dominant gene with a similar phenotypic effect?

9. What configuration of salivary gland chromosomes would you expect with complete conjugation in a Drosophila female with one closed X chromosome and one open homolog with an inversion involving its middle third?

10. In such a female as described in the preceding problem, what meiotic products would be expected following single exchange within the inverted segment of the X chromosomes? What genotypes would be expected among the offspring of such a female mated to a normal male?

CHAPTER X

LETHALS

As shown in Chapter IX, individuals in which a specific section of a chromosome is absent usually fail to develop—that is, the particular gene-combination present in their cells is lethal to the individual. There are many cases known in which specific types of homozygotes fail to develop and in which no structural derangement of the chromosomal material can be detected. In such instances we attribute the failure to develop to the action, or perhaps lack of action, of a particular allele. Obviously there is a real difficulty in differentiating between very short deficiencies and reorganizations within the gene itself. This difficulty is considered further in Chapter XIII. For purposes of the discussion in this chapter, we shall assume that there are intragenic changes that, under certain conditions, lead to death of the organism. We commonly refer to genes modified in this way as "lethal genes," an expression justified solely by its convenience. It may be noted at once that the lethal genes ordinarily studied are recessive; a gene with a dominant lethal effect is necessarily lost before it can be studied.

White Seedlings in Maize.—Many recessive genes are known in maize which, in homozygous condition, influence the zygote in some manner such that chlorophyll does not develop. This is an example of many different loci being concerned with a single end result, in this instance with the production of chlorophyll. Plants lacking chlorophyll, known as white seedlings (gene symbols: w_1, w_2, etc.), develop for several days at the expense of food material stored in the endosperm, just as normal green seedling plants do. This food is usually sufficient to enable them to develop two seedling leaves of normal size. If we grow seeds produced by self-pollinating a plant heterozygous for a single white seedling gene, and classify the resulting seedlings early in development, a standard 3:1 ratio of normal to white seedlings will be obtained. If, however, we classify the plants in these same cultures several weeks later, there will be no white plants and the ratio will be 1:0. The white plants are unable to survive after the stored food reserves are exhausted. If the surviving green plants are allowed to mature and their genotypes determined by self-pollinating them and growing seedling cultures, it will be found that two-thirds of them are heterozygous for the w gene and one-third are homozygous for its normal allele. In other words, inheritance of this character follows exactly the same principles as does that of an ordinary non-lethal recessive character—the only difference is that the homozygous recessive type does not survive. White seedling genes are known in a great many different plants.

Time of Death of Inviable Zygotes.—Homozygous types may fail to survive for many different reasons. For example, recessive genes for germless seeds are known in maize; in these cases the genes concerned result in failure of the embryo to develop, but have no apparent influence on the endosperm. Other genes result in the opposite condition, development of the embryo and failure of the endosperm. Still others are known which result in failure of both endosperm and embryo—no seed at all develops. This latter class represents a type of effect that is very common, particularly in animals, namely one in which a homozygous class dies before it is convenient to classify the zygotes. Lethals of this type can be worked with by special techniques, as we shall see.

The Yellow Mouse.—As a special case of lethals in which the zygotes are not readily classifiable, we may consider the now classical yellow mouse case, first studied by Cuénot. Yellow is dominant to non-yellow. If, for example, a yellow mouse is crossed with a black mouse, there will be two types in F_1, yellow and black in the ratio 1:1. If yellow mice of this generation are inter-bred, the F_2 mice will segregate so as to give 2 yellows to 1 black. Now, if the yellow mice are tested individually for genotype by crossing them with black mice, it will be found that not one of them will be homozygous for yellow. In other words, it is not possible to have a homozygous stock of yellow mice. Castle and Little explained the inheritance of the yellow character by assuming that the homozygous yellow genotype is lethal. This has been checked by examining unborn embryos from yellow parents. It is found that approximately one-fourth of the embryos from such matings do die in the uterus, as would be expected from the hypothesis. This type of case, in which the heterozygote is visibly different from the homozygous wild-type, and in which the homo-

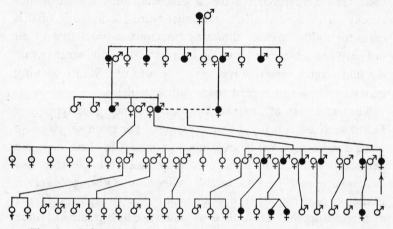

Fig. 63.—Pedigree showing the inheritance of brachyphalangy, a dominant skeletal abnormality in man. In the third generation a first cousin marriage of affected individuals is shown by dotted lines. The daughter of this marriage, designated with the arrow, is the presumed homozygous type discussed in the text. Note affected identical twins in the last generation shown. (From Mohr and Wriedt—part of pedigree omitted because of lack of space.)

zygous mutant type is lethal, is rather common in animals and is known in some plants. Many dominant characters in Drosophila are of this type, for example, the spread wing and bristle character dichaete (D), the character hairless (H), the rough and small eye character known as glued (Gl), the minutes (M), and many others. In all of these instances the visible character is treated as a dominant although, of course, the lethal effect of the gene concerned is recessive.

Brachyphalangy.—A number of abnormalities in man appear

to be inherited as simple dominant characters.   Examples of this
are brachydactyly (short fingers and toes), brachyphalangy (shortening of second phalanx of second fingers and toes), polydactyly
(extra fingers and toes), and others.   The case of brachyphalangy,
studied by Mohr and Wriedt, is particularly interesting because
one individual presumably homozygous for the character is
recorded.   In the rather extensive pedigree shown in figure 63,
there is indicated in the third generation a first cousin marriage
of brachyphalangous individuals.   Of the two recorded daughters,
one showed the abnormality in typical form; the other was without fingers and toes and showed general skeletal abnormalities.
This cripple lived for only a year.   Although the proof is not
complete, the presumption is strong that this individual was
homozygous for the gene for brachyphalangy.   By analogy with
dominant characters in Drosophila, it is probable that a number
of the other known dominant characters in man would be lethal
in the homozygous form.   Lethals of various kinds may account
for a proportion of the still-births in man.

Semi-lethals.—As will be emphasized in Chapter XIII (Mutation), there are all intermediate conditions between those in
which a mutant gene has no detectable effect on viability of
the homozygote and those in which the homozygote dies at a
very early stage.   Near the lethal end of this range of variation are to be found the *semi-lethal* genes, which usually result
in early death.   It should be emphasized that this classification
is primarily one of convenience.   As has already been pointed
out, the ratio obtained may depend on the stage of development
at which individuals are scored; death may occur at any stage
after fertilization.   Sometimes the adult is short-lived.   As a
matter of fact, very few of the standard mutant types with which
geneticists work are equal in viability to the wild-type—most of
them would be much less viable than they are were nearly
optimum culture conditions not maintained.

Sex-linked Lethals.—Hemophilia in man is an example of a
semi-lethal character.   Affected males usually do not live to reproductive age and even when they do, their life-expectancy is much

lower than that of non-bleeders. This particular semi-lethal character happens to be sex-linked. Many sex-linked lethals and semi-lethals are known in Drosophila. Sex-linked lethal genes can be transmitted by females only. If a female is heterozygous for a lethal gene, all her sons that receive the lethal-bearing X chromosome die at an early stage of development. The sex ratio in the adult progeny is modified from 1:1 to 2:1 as indicated by the following scheme:

$$+/l \; ♀ \times + \; ♂$$

$$1 \; +/+ \; ♀ \; ; \; 1 \; +/l \; ♀ \; : \; 1 \; + \; ♂ \; ; \; 1 \; l \; ♂ \; \text{(dies)}$$

If a non-lethal sex-linked gene and a sex-linked lethal are segregating in the same individual, the ratio for the non-lethal character will be distorted. The amount of the distortion will depend of course on the amount of crossing over between the two loci. With a lethal in the same chromosome as yellow and 10 units away we would expect the following results:

$$\frac{+ \quad +}{y \quad l} \; ♀ \times \underline{y \quad + } \; ♂$$

(Females)   (Males)

$$9 \; \frac{+ \quad +}{y \quad +} \qquad\qquad 9 \; \underline{+ \quad +}$$

$$9 \; \frac{y \quad l}{y \quad +} \qquad\qquad 9 \; \underline{y \qquad l} \; \text{(dies)}$$

$$1 \; \frac{+ \quad l}{y \quad +} \qquad\qquad 1 \; \underline{+ \qquad l} \; \text{(dies)}$$

$$1 \; \frac{y \quad +}{y \quad +} \qquad\qquad 1 \; \underline{y \qquad +}$$

In this particular cross the ratio for yellow in the daughters is 1:1 but in the sons is 9 + to 1y.

Effects of Lethals on Ratios.—In the case of autosomal lethals in animals the usual method of detection is through their effect on ratios for visible characters. The direction and amount of the distortion in a ratio for a visible character will depend on the arrangement of genes in the chromosomes and on the amount of crossing over. This is illustrated by a hypothetical example in Drosophila, in which the gene a and the lethal gene l are in separate members of a homologous pair of chromosomes and are 10 units apart (Fig. 64).

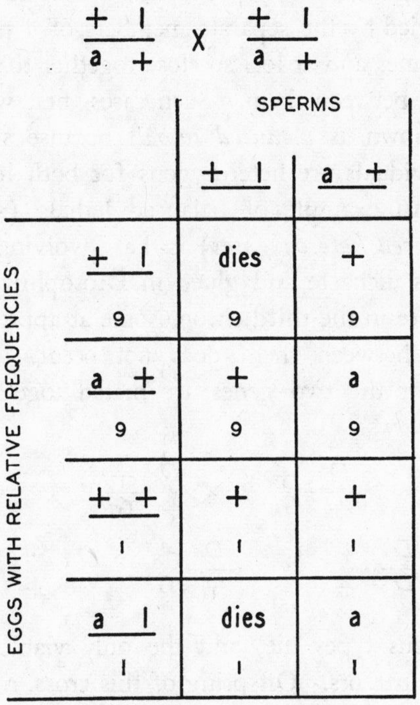

$$\frac{+\quad l}{a\quad +} \;\; X \;\; \frac{+\quad l}{a\quad +}$$

	SPERMS	
	$\dfrac{+\quad l}{}$	$\dfrac{a\quad +}{}$
$\dfrac{+\quad l}{}$ 9	dies 9	+ 9
$\dfrac{a\quad +}{}$ 9	+ 9	a 9
$\dfrac{+\quad +}{}$ 1	+ 1	+ 1
$\dfrac{a\quad l}{}$ 1	dies 1	a 1

EGGS WITH RELATIVE FREQUENCIES

Fig. 64.—Checkerboard showing the effect of a lethal on the genetic ratio of a recessive character differentiated by a linked gene. A crossover value of 10 units between the lethal (l) and the recessive non-lethal gene (a) is assumed.

Collecting viable phenotypes, we have:

$$\begin{array}{ll} +\quad + & 20 \\ a\quad + & 10 \end{array}$$

or a 2:1 ratio. This ratio is independent of the amount of crossing over in the female. If a recessive non-lethal gene and a lethal gene are carried in the same member of a pair of homologs, then the phenotypic ratio will depend on the amount of crossing over between the two loci.   It is possible to determine the map position of lethal gene loci by the use of methods of this kind.

BALANCED LETHALS

An interesting and useful special case is that in which different lethals are carried by the separate members of a pair of homologous chromosomes and at loci so close together that crossing over does not occur between them.   Such cases, first worked with by Muller, are known as *balanced lethals* because such a stock in which all individuals are heterozygous for both lethals will still breed true.   An example of balanced lethals (sometimes also known as *balanced heterozygotes*) is that involving the two dominant characters dichaete and glued in Drosophila.   Both the *D* and *Gl* genes are in the third chromosome at approximately 41.0.   Crossing over between them does not occur.   If individuals heterozygous for the two genes are mated together results are obtained as follows:

$$\frac{D \quad +}{+ \quad Gl} \, ♀ \times \frac{D \quad +}{+ \quad Gl} \, ♂$$

$$\text{I} \,\, \frac{D \quad +}{D \quad +} \quad ; \quad 2 \,\, \frac{D \quad +}{+ \quad Gl} \quad ; \quad \text{I} \,\, \frac{+ \quad Gl}{+ \quad Gl}$$

Both homozygous types die, and the only viable type is genetically like the parents.   Offspring of this cross, mated together, repeat this behavior.   In this way it is possible to maintain indefinitely a stock in which all viable individuals are phenotypically dichaete glued.   If a lethal gene were known at locus 41.0 that had no detectable phenotypic effect in heterozygous condition, it would be possible to balance either dichaete or glued with such a recessive lethal.   Any crossing over between the two loci at which lethal genes are located will result in the breakdown of the balanced system.   In practice it rarely happens that two non-allelic lethals are so close together that they never or only rarely show crossing over.   Even if they are further apart, however, it is often possible to make up a balanced stock by making use of an inversion which prevents effective crossing over between the two loci.

For example, the dominant gene hairless (H), which is lethal when homozygous, can be balanced by using an inversion known as "In 3RP" (an inversion in the right limb of the third chromosome). This particular inversion is lethal when homozygous, that is, carries a recessive lethal. Such a balanced stock is represented as follows:

$$\frac{H+}{(In\ 3\ RP)\quad +\quad l}\ \male \times \frac{H+}{(In\ 3RP)\quad +\quad l}\ \female$$

$$1\ \frac{H\quad +}{H\quad +}\ ;\ 2\ \frac{H\quad +}{(In\ 3RP)\quad +\quad l}\ ;\ 1\ \frac{(In\ 3RP)\quad +\quad l}{(In\ 3RP)\quad +\quad l}$$

Again only the heterozygous type survives. Still other combinations are possible. Both lethals may be without phenotypic effect in the heterozygote and may be used to balance a non-lethal gene in the heterozygous condition, for example, as in the general case,

$$\frac{a\quad l_1\quad +}{In\quad +\quad +\quad l_2}.$$

Balanced Lethals in Keeping Stocks.—Balanced lethals are extremely useful in maintaining characters that cannot be kept in homozygous condition. The dichaete character, for example, is inconvenient to keep in the unbalanced condition; dichaete flies must be selected each generation to prevent the stock from becoming homozygous for the wild-type allele. Glued is another character of the same type. Combining them produces a balanced true-breeding stock which requires no selection. On outcrossing a D/Gl balanced stock to wild-type, two phenotypes are obtained, $D/+$ and $Gl/+$, neither of which will breed true.

Sterile Types.—In addition to the use of balanced lethals in keeping characters that are completely lethal when homozygous, they are also made use of in keeping types that are viable but sterile and types that have such low viability that it would be difficult to maintain them in homozygous condition. As a special case involving the X chromosome where a true balanced lethal condition is not possible, we may consider the standard method of keeping the singed character. Singed (sn) is a sex-linked recessive affecting bristle form. Homozygous singed females are sterile, since they lay eggs that fail to hatch, regardless of the sperm with which they are fertilized. Singed males, on the other hand, are fully fertile. There is an X chromosome inversion containing, among other genes, a recessive lethal and a *B* (bar) gene. There is practically no effective crossing over between a normal chromosome and a chromosome carrying this inversion, known as "*ClB*" (read "cee el bee"—the "*C*" is a relic symbol for "crossover reducer"). Females heterozygous for *sn* and *ClB* mated to *sn* males give types as follows:

$$\frac{ClB}{sn} \; ♀ \; \times \; sn \; ♂$$

$$1 \; \frac{ClB}{sn} \, ♀ \; ; \; 1 \; \frac{sn}{sn} \, ♀ \; ; \; 1 \; ClB \; ♂ \; ; \; 1 \; sn \; ♂$$

(Note that we have here omitted all designation of normal alleles. This is a convenient and common practice applicable wherever confusion does not result.) The homozygous singed females are sterile and the *ClB* males die. The only types that breed are like the parents and the stock is therefore self-perpetuating.

It is evident that self-perpetuating stocks of singed flies can be maintained by the use of attached-X chromosomes. If an attached-X female of any constitution is mated to singed males, all daughters will be like the mother with respect to sex-linked genes and all the sons will be singed.

In addition to singed, there is a large class of genes that result in sterility in one or both sexes. Singed is such a character in Drosophila. There is a character known as *female sterile* in Drosophila differentiated from normal by a second chromosome recessive. Homozygous females have rudimentary ovaries but are externally normal in appearance. Homozygous males are fertile. This gene is maintained in a balanced condition. Other examples could be cited in Drosophila and in other animals. Similar types are known in many plants. In maize there are several recessive characters in which the pistillate inflorescence is absent or produces no functional female gametophytes. All such genes that result in sterility are non-lethal to the individual but are effectively lethal from the standpoint of the species.

Lethals and the Diploid Condition.—The often-used analogy between a living organism and a watch is useful in thinking of lethals. Almost any random change made in a watch is bad for the watch. If the change is extensive, removal of a wheel for example, the result is that the watch does not run—a change analogous to a lethal. Moving the hair spring regulator at random is very likely to result in a watch that runs but does not run at the proper speed—a change analogous to a non-lethal "visible" character. Of course the diploid organism has one advantage over the watch—it has two sets of homologous chromosomes which in a sense work on the principle of double assurance—usually a lethal gene must be present in duplicate to bring about a lethal end result. It is as though a watch were made in such a way that every part were present in duplicate, so that removal of one wheel of a specific type would not stop the watch—the homologous wheel, too, would have to be removed.

REFERENCES

Bridges, C. B., and T. H. Morgan. 1923. The third chromosome group of mutant characters of Drosophila melanogaster. Carnegie Inst. Washington, publ., 278:123–304.

Chesley, P. 1935. Development of the short-tailed mutant in the house-mouse. Journ. Exper. Zool., 70:429–459.

Demerec, M. 1934. Biological action of small deficiencies of X chromosome of Drosophila melanogaster. Proc. Nat. Acad. Sci., 20:354–359.

Mohr, O. L. 1929. Letalfaktoren bei Haustieren. Züchtungskunde, 4:105–125.

PROBLEMS

1. A zygotic lethal gene (zl) is linked with the pericarp color gene (P) in maize and gives 2 per cent crossing over. What will be the ratio of P to $+$ in progeny obtained by self-pollinating $\dfrac{P \quad +}{+ \quad zl}$ plants?

2. Assuming a recessive gene for germless seeds (gm) to be linked with the gene for liguleless (lg) in maize, what ratio of $+$ to lg plants would be expected on self-pollinating $\dfrac{+ \quad +}{lg \quad gm}$ plants with (a) no crossing over, (b) 10 per cent crossing

over, and (c) 40 per cent crossing over? What ratio of $+$ to lg plants would be expected from $\dfrac{+\quad gm}{lg\quad +}$ plants with these same three frequencies of crossing over?

3. How could one make use of attached-X chromosomes in Drosophila for keeping the lozenge character (homozygous females sterile, males fertile) in a stock that would require no selection?

4. Inversion scute-7 exists in the same chromosome with the recessive allele of singed (sn). As indicated in the text, crossing over does occur between the inversion end and the singed locus, but this is so low that it can be disregarded for stock-keeping purposes. Knowing that homozygous singed females are sterile, how could you use the In $scute$-7 sn stock in keeping a notch deficiency chromosome? In $scute$-7 includes the segment corresponding to that for which notch is deficient.

5. Assuming that l is a lethal in the third chromosome of Drosophila, what ratio of Dichaete to wild-type would result from the cross $\dfrac{D\quad l}{+\quad +}\ ♀ \times \dfrac{D\quad +}{+\quad l}\ ♂$ with no crossing over between the lethal and the dichaete locus? What ratio would be expected with 10 per cent crossing over between the lethal and dichaete?

6. If a and b, recessive genes for female and male sterility respectively, are linked and give no crossing over with each other, what types will result from the cross $\dfrac{a\quad +}{+\quad b}\ ♀ \times \dfrac{a\quad +}{+\quad b}\ ♂$? If these interbreed at random what types of matings will contribute to the next generation and what genotypes will they give?

7. If l_1 and l_2 are non-allelic sex-linked lethals in Drosophila that do not cross over with each other, what sex-ratio would the cross $\dfrac{+\quad l_2}{l_1\quad +}\ ♀ \times +\ +\ ♂$ give? (Such a female can be obtained by "covering" one lethal in the male with a short duplication or in other ways, as will be seen in the next chapter.)

8. There is an allele (bb^l) of the bobbed gene in Drosophila that is lethal in the female when homozygous. What will be the sex-ratio among the offspring of the cross $\dfrac{bb^l}{+} \times bb^l$? Remember that the Y chromosome carries the normal allele of bobbed.

9. If a female, supposedly of the constitution $\dfrac{+\quad v\quad +}{ct\quad +\quad g}$, when crossed to a vermilion male, gave offspring of the following phenotypes:

Females

+	+	+	420
+	v	+	430

Males

+	v	+	318
ct	v	+	54
+	v	g	37
ct	v	g	2

What could you deduce as to the true genotype of the original female?

10. Look up in the previous chapter the products of single crossing over between the two X chromosome inversions, *In scute-8* and *In scute-4*. The bobbed-deficiency scute-duplication chromosome can be carried in males, which are fertile; the other crossover chromosome can be carried in heterozygous condition in females. If the two chromosomes are put in the same female, what will be the products of crossing over between them?

CHAPTER XI

TRANSLOCATIONS

ONE type of chromosome rearrangement that may result from illegitimate crossing over (Fig. 46) is that in which segments of non-homologous chromosomes are interchanged. Such reciprocal translocations are known in many organisms, both plant and animal. Rearrangements of this kind occasionally occur spontaneously; their frequency can be greatly increased by x-ray or radium treatment. In several species of seed-plants it is found that wild populations may contain chromosome sets that differ from one another by one or more reciprocal translocations. It is also clear that related species sometimes differ in this respect, and that the phenomenon of translocation has been of importance in evolutionary divergence within groups.

Conjugation in Translocation Heterozygotes.—Reciprocal translocations may be homozygous or heterozygous. In the latter case—in which one set of chromosomes has the old and the other set has the new arrangement of segments—the individual is often known as a *translocation heterozygote.* We shall consider first the behavior (in plants) of a translocation heterozygote in which the chromosomes involved are approximately equal-armed (centromere near the middle), and in which the translocation has interchanged large segments—approximately whole arms. During pachytene corresponding segments of chromosomes associate in pairs, *i.e.,* conjugate. Complete regionally specific conjugation necessitates an exchange of pairing partners at the point of change in homology (Fig. 65). This change of partners provides a method of locating the translocation points, and has been used extensively in this way in studies of reciprocal translocations in maize.

Chiasmata in Translocation Heterozygotes.—Within each of the arms of the pachytene cross, chiasmata are formed as for the corresponding segments in a structural homozygote (Fig. 65). During diplotene there may be more or less terminalization of

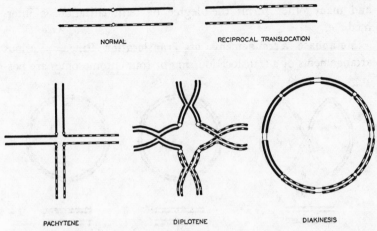

NORMAL

RECIPROCAL TRANSLOCATION

PACHYTENE

DIPLOTENE

DIAKINESIS

Fig. 65.—Scheme showing conjugation and chromosome association at later stages for a simple reciprocal translocation heterozygote. Above is shown one representative of each of the two chromosomes involved, at the left, in the normal condition and, at the right, after exchange of arms. Terminalization is assumed to be complete at diakinesis.

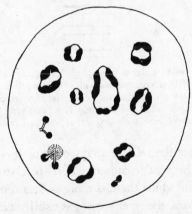

Fig. 66.—Diakinesis showing a reciprocal translocation ring of four chromosomes in Datura. Terminalization of chiasmata is complete here. (After Belling.)

chiasmata, depending on the species. In Datura (jimson weed), in Oenothera (evening primrose), and in Campanula, terminalization is more or less complete, and the translocation complex forms an open ring of the type shown in figure 66. In Pisum (garden pea) there is little or no terminalization, while in maize and many other forms the degree of terminalization is intermediate.

Metaphase Arrangements of Translocation Rings.—Various arrangements of a translocation ring of four chromosomes are pos-

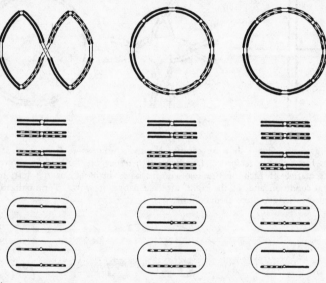

Fig. 67.—Representing different metaphase arrangements at the first meiotic division of a ring of four chromosomes in a translocation heterozygote. The immediate products are shown directly below the arrangement concerned, and the final products are shown at the bottom. Note that only those gametes from the leftmost arrangement have each segment represented once and only once.

sible at metaphase; the relative frequencies of these are important in determining the genetic consequences. In Datura, Campanula, and Oenothera, in which there is more or less complete terminalization of chiasmata, the arrangement is usually such that alternate members of the ring go to the same pole at the first division (Fig. 67). This type of disjunction gives gametes like the parents of the translocation heterozygote, with respect to the arrangement of segments. Two other arrangements are possible, in both of which adjacent members of the ring go to the same pole at the first division (Figs. 67 and 68).

Fig. 68.—Two metaphase arrangements in a translocation ring of four chromosomes in maize. In the left figure alternate members of the ring are oriented toward the same spindle pole while in the right figure adjacent members of the ring are arranged so as to go to the same spindle pole. (From Darlington.)

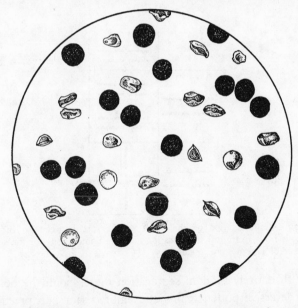

Fig. 69.—Sample of pollen of a maize plant heterozygous for a reciprocal translocation. The light shrivelled grains contain little or no starch and are incapable of functioning. (Drawn by E. Scott.)

Disjunction of Translocation Rings.—Random two-by-two separation of the four members of the translocation ring will give six types of spores (gametes in animals) in equal numbers.

Four of these six types will have one segment present in duplicate and another segment not represented at all (Fig. 67). Ordinarily in plants these duplication-deficiency spores do not give rise to viable gametophytes. For example, such microspores in maize usually give rise to defective pollen grains which contain little or no starch and are incapable of functioning (Fig. 69). Thus, one result of the heterozygous translocation condition is partial steril-

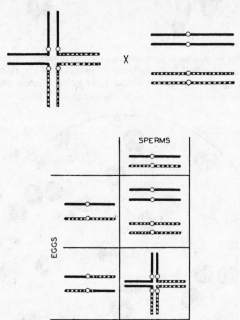

Fig. 70.—Representation of the results of a cross between an individual heterozygous for a reciprocal translocation and one homozygous for the standard arrangement of chromosome segments.

ity. With entirely random segregation there should be 66.7 per cent of defective pollen grains and embryo sacs. In practice, in maize and other plants, the sterility is usually nearer to 50 per cent. Two factors are of importance in bringing about this reduction: (1) exchange in the region between the translocation point and the centromere, and (2) reduction in frequency of the metaphase arrangements in which adjacent chromosomes go to one pole.

Partial Sterility and Identification of Translocation Hetero-zygotes.—The partial sterility characteristic of a translocation heterozygote is made use of in detecting new occurrences of trans-locations in the first instance (not *all* partial sterility is due to this cause, however), and in working with them after they are detected. Using the method of pollen sterility, Anderson and others have obtained translocations involving many different com-

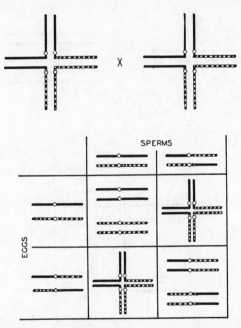

Fig. 71.—Results of self-fertilizing a translocation heterozygote, or of crossing two individuals heterozygous for the same translocation. Note the new type homozygous for the translocation.

binations of two of the ten chromosomes of maize. The inherit-ance of a translocation can be followed in essentially the same way as can that of a dominant gene, by using partial sterility as an index of heterozygosis. This will be more obvious after we have described the offspring produced by translocation heterozygotes.

Translocation Heterozygotes.—In plants, in which the duplica-tion-deficiency products of meiosis are eliminated in the gametophyte, a cross between a translocation heterozygote and a strain homozygous for the original arrangement (frequently referred to as the "standard" arrangement), will result in two types in a 1:1 ratio, as shown in figure 70. One of these will be homozygous and like the standard. The other will be a translocation heterozygote showing partial pollen and embryo sac sterility. If a translocation heterozygote is self-pollinated, or if two similar

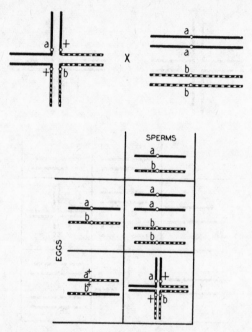

Fig. 72.—Showing linkage of loci, originally in non-homologous chromosomes, in a translocation heterozygote. There is assumed to be no crossing over between the loci involved and the translocation breaks.

heterozygotes are crossed, there will result three types in the ratio 1:2:1 (Fig. 71). In addition to the homozygous standard type and the heterozygote, there will be a new homozygous type in which both sets of chromosomes have the new arrangement; that is, the new type is homozygous for the translocation. Cytologically this homozygote is essentially like the original normal. It has only pairs of chromosomes at meiosis.

Translocations and Linkage Groups.—Obviously an interchange between non-homologous chromosomes will change the linkage relations of the genes carried in the chromosomes involved. A translocation heterozygote will show linkage between genes that would normally segregate independently, as is shown in figure 72. An interesting characteristic of linkage in a translocation heterozygote is that a linear map will not be obtained for genes in the chromosome pairs involved in the translocations, but a 4-armed

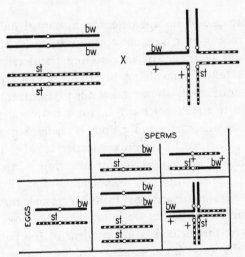

Fig. 73.—Scheme showing how complete linkage of brown (*bw*—second chromosome) and scarlet (*st*—third chromosome) can result in the male of Drosophila from a reciprocal translocation. The original arrangement of genes is indicated in the female which is homozygous for the normal arrangement of chromosome segments.

map corresponding to the pachytene configuration. A translocation homozygote will differ from the standard in its linkage groups just as expected; that is, there will be segments of the chromosome maps interchanged, corresponding to the interchanged physical segments. Thus if the original linkage groups are A B C D and E F G H, the new ones may be A B G H and E F C D.

Translocation Heterozygotes in Drosophila.—Translocations in Drosophila have been extensively studied. In this organism, unfortunately, it has not been possible to observe in any detail the meiotic behavior of such rearrangements. The pairing properties in translocation heterozygotes can be studied in salivary gland chromosomes, and it is found that here the behavior is essentially the same as in pachytene in plants. Corresponding segments pair, and there is a change of partners at the point of change in homology. The exact alignment of corresponding bands makes it relatively simple to determine the exact points of interchange in terms of bands.

As in the case of linkage detection in normal individuals, the fact that crossing over does not occur in the males of Drosophila can be taken advantage of in detecting translocations. As an example, a testcross involving the two eye color genes brown (*bw*) and scarlet (*st*) shows independent assortment in standard stocks. But if a male heterozygous for a reciprocal translocation involving chromosomes 2 and 3, and also heterozygous for brown and scarlet, is crossed to a brown scarlet female, there will be only two instead of four classes of offspring. In other words, brown and scarlet will show complete linkage (Fig. 73). In this case the breaks are arranged in such a manner that the normal alleles of brown and scarlet are in one new chromosome. This is not necessary; the method is independent of the position of the breaks in the two chromosomes. Table 1 summarizes actual results obtained by Dobzhansky in an experiment in which males were x-rayed. It has been found that translocation in Drosophila may involve any two of the four chromosome pairs, *i.e.*, X–2, X–3, X–4, Y–2, Y–3, Y–4, 2–3, 2–4, and 3–4.

In female translocation heterozygotes in Drosophila, crossing over occurs in all segments. It is found, however, that crossing over near the translocation breaks is reduced in many cases. The extent of this reduction in crossing over varies, depending on the position in the chromosome of the translocation break, and is presumably the result of mechanical difficulties in conjugation. Near the breaks in a heterozygous translocation it is often observed cytologically that pairing is incomplete. There is a tendency for segments near the translocation breaks to compete for pairing

TABLE I

Results from Dobzhansky illustrating a genetic method of detecting chromosome translocations in Drosophila. Bristle (*Bl*—second chromosome) dichaete (*D*—third chromosome) males were *x*-rayed and mated to attached-X females homozygous for the fourth chromosome recessive eyeless (*ey*). The bristle dichaete sons of this mating were crossed to eyeless females. The classification for sex is omitted in this table. The counts from 112 normal cultures are presented together; in each of these all of the eight phenotypic classes were represented. The final mating from which the recorded offspring were obtained was therefore $+ + ey \,♀ \times Bl/+ \; D/+ \; +/ey \,♂$.

Cultures	Phenotypes								Translocation indicated
	Bl D +	*Bl D ey*	*Bl + +*	*Bl + ey*	*+ D +*	*+ D ey*	*+ + +*	*+ + ey*	
112 "normal" cultures	2324	1991	2183	2140	2246	1897	2196	2067	none
1	46	36	....	....	....	....	56	50	2-3
2	49	33	....	....	....	....	38	40	2-3
3	23	15	....	....	....	....	9	12	2-3
4	6	12	....	....	....	....	11	14	2-3
5	37	....	....	22	30	....	....	33	3-4
6	34	....	....	19	27	....	....	35	3-4
7	41	....	....	43	34	....	....	22	3-4
8	21	....	....	29	20	....	....	26	3-4
9	34	....	....	28	43	....	....	27	3-4

partners. This concept of competitive conjugation has been developed especially by Dobzhansky. Quantitative studies of crossing over in plants are more difficult than in Drosophila, and relations of this kind are consequently not as well understood.

Translocation Homozygotes in Drosophila.—Translocation homozygotes in Drosophila, when they are obtained, show essentially the same behavior as do those of plants. It turns out, however, that a high proportion of the translocations known in Drosophila are lethal when homozygous, whereas of those known in maize and in other plants, very few or none are lethal. This difference, somewhat surprising at first thought, has a relatively simple explanation. In plants, translocations must be transmitted through the gametophyte generations in order to be transmitted at all. Those that would be lethal when homozygous, then, are presumably lethal in the gametophyte and are eliminated before they can be studied.

Duplication-deficiency Gametes in Drosophila.—As implied above, translocations that are lethal in Drosophila can be transmitted through both eggs and sperm. This is not surprising,

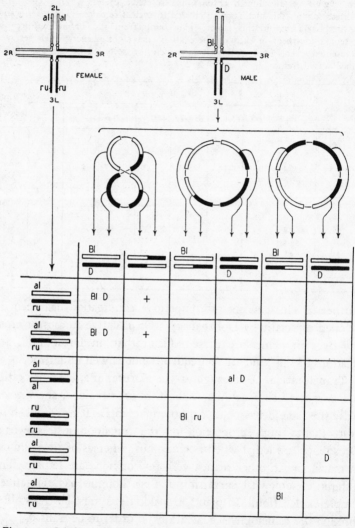

Fig. 74.—Scheme showing method of demonstrating that duplication-deficiency gametes are functional in Drosophila. In all cases the position of the chromosomes in the gametes are such that right and left limbs, as indicated above by 2L, 2R, 3L, and 3R, are properly oriented. The second chromosome

since we have already pointed out that eggs with no X chromosome at all can function. In fact Y-bearing sperm, which are normally produced, are deficient in almost all of the X chromosome genes. Studies of the offspring of two translocation heterozygotes in Drosophila show directly that types which would not survive in the gametophyte in a plant, function perfectly well in both eggs and sperm. As figure 74 shows, viable zygotes can be obtained from the union of duplication-deficiency gametes, providing the two gametes concerned carry complementary segregation products; that is, if the two gametes mutually compensate for deficiencies and duplications. The inviability of uncompensated zygotes has been established by determining the amount of egg mortality.

Other Types of Translocations.—Numerous deviations from the behavior of a translocation heterozygote described above are possible. For example, cases are known in maize in which two very small segments are interchanged. Heterozygotes of this type usually show two pairs of chromosomes at meiosis, presumably because of the lack of chiasma formation between the short terminal segments. If a long and a short segment are interchanged, the result is usually a chain of. four chromosomes rather than a ring of four. Even with large segments interchanged, occasional cells will be found in which chiasmata fail to form in one or more arms of the pachytene cross and a univalent plus a chain of three, two pairs, or a pair plus two univalents may result. Segregation is not always regular; three chromosomes of the ring may go to one pole, and only one to the other. This may lead to the formation of zygotes with entire extra chromosomes. Such extra chromosomal types will be discussed in Chapter XV.

---

is shown in outline, the third in black. In the diagrammatic representations of metaphase arrangements only half the true number of chromatids is shown—the omitted ones are exact duplicates of those shown. Rectangles with no phenotypes indicated, represent zygotes that do not survive. The phenotypes *Bl ru*, *al D*, *Bl*, and *D* can only arise from the union of complementary duplication-deficiency gametes. It is possible, by making use of more marker genes, to differentiate the four types arising from balanced gametes. The precise method indicated has not been used in practice, but Dobzhansky has used one essentially similar.

Translocations in Drosophila involving short segments may give rise to viable zygotes in which one of the interchanged segments is deficient and the other present in excess (Fig. 75). Duplications or deficiencies obtained in this way have essentially the same properties as those obtained in other ways (Chapter IX).

Of the known translocations, by far the majority are reciprocal. Apparent instances of *simple translocations* in which a segment of one chromosome is removed and attached to an unbroken non-homologous chromosome have been described, but none of these

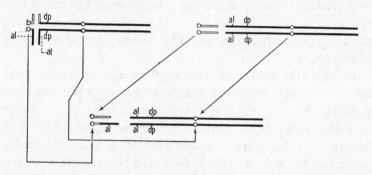

Fig. 75.—Method of origin of a duplication type in Drosophila from a cross of a 2-4 translocation heterozygote and a normal fly. The second chromosome is shown in black, the fourth in outline. A fly deficient in a segment of one fourth chromosome may be viable. By making use of triplo-4 flies, or in other ways, it is possible to get a duplication carrying individual like that shown but with one more fourth chromosome, *i.e.*, with two entire chromosomes in addition to part of one. (Based on data of Dobzhansky.)

cases has been established with certainty. If simple translocations occur at all, they are very infrequent relative to reciprocal ones.

Cytological Demonstration of Crossing Over.—Creighton and McClintock have made use of a translocation as a physical "marker" in the chromosome in studying the visible results of crossing over. In a translocation heterozygote in maize, also heterozygous for a terminal "knob" on one of the chromosomes involved, it was possible to show cytologically that recombinations for translocation and knob were produced at meiosis and, furthermore, that this was correlated with genetic crossing over between genes known to lie in the segment between the translocation and the knob (Fig. 76). A similar cytological demonstration of crossing over involving a larger number of individuals was made by Stern in Drosophila, where a more precise control of genetic crossing over was possible (Fig. 77).

Cytological Maps.—As indicated above, it is possible to localize translocation breakage points both genetically and cytologically.

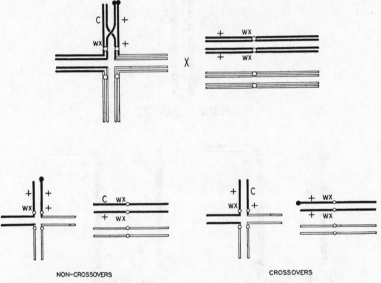

NON-CROSSOVERS   CROSSOVERS

Fig. 76.—Diagram of the type of cross used in maize in demonstrating cytological crossing over and its correlation with genetic crossing over. The knob indicated on the *C wx* arm of chromosome 9 (black) is presumably made up of heterochromatin and is readily observed cytologically. The other chromosome involved in the actual experiment was number 8 (shown in outline). In this figure no attempt is made to indicate the actual relative lengths of the arms of the two chromosomes involved. (Based on work by Creighton and McClintock.)

It was previously pointed out that similar correlations can be made by using inversions, deficiencies, and duplications. The use of these methods has made it possible to localize many genes in the physical chromosomes, and in this way to construct what have been called *cytological maps*. Before the discovery of salivary gland chromosomes, such cytological maps were based on somatic metaphase chromosomes. For this purpose translocations involving an exchange of a large, easily visible, segment from one chromosome and a very small segment, the size of which may be more or less disregarded, from a second chromosome are useful. Such a cytological map is shown in figure 78. Salivary gland

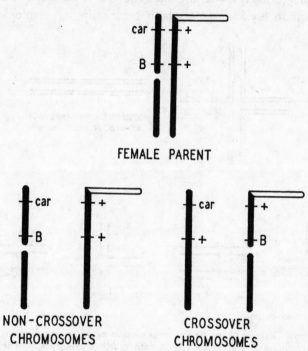

FEMALE PARENT

NON-CROSSOVER
CHROMOSOMES

CROSSOVER
CHROMOSOMES

Fig. 77.—Diagram of method used by Stern for detecting crossing over cytologically. A segment of the Y chromosome, shown in outline, is attached to the X chromosome at the centromere. One of the X chromosomes of the female parent is represented as broken. The proximal segment has its own centromere and the distal segment is attached to the fourth chromosome centromere; *i.e.*, there is an X-4 translocation. The offspring of females of this constitution, mated to carnation males, were classified for the two characters indicated. Cytological examinations showed that the constitutions with respect to these two genes were correlated with the structure of the X chromosome as indicated in the diagram.

chromosome maps can be made very much more precise, since it is possible to localize genes to within the limits of one or two cross bands. Salivary gland chromosomes, however, have one disadvantage as compared with somatic chromosomes—they do not give a clear picture of inert material. A partial cytological map of the X chromosome is given in figure 48. It is evident from an examination of these cytological maps that crossing over per unit distance is far from constant. This is shown by a clumping of genes near the centromeres in the crossing over maps of chromosomes 2 and 3, and near the ends of chromosomes 1 and 2. Actually, the genes appear to be distributed more or less at random

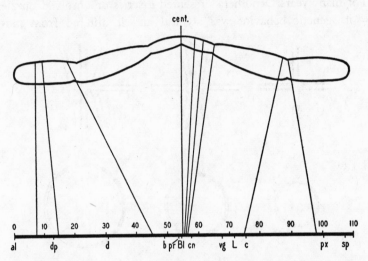

Fig. 78.—Cytological map of the second chromosome of Drosophila melanogaster. The genetic map, based on crossing over, is shown below. The vertical lines connect this with corresponding points in the metaphase chromosome shown above. cent.—centromere. (After Dobzhansky.)

in the cytological chromosomes. A notable exception to this even distribution is the inert region of the X chromosome (proximal third) which contains only one known gene.

In the same way that cytological maps can be made in Drosophila, they are possible in maize. In this case correlations are most easily made with pachytene chromosomes. Although a number of genes have been localized within the physical chromosomes in this plant, the cytological maps are not as complete as are those for Drosophila.

OENOTHERA

With a knowledge of translocations, we are prepared to consider the special case of the evening primrose (Oenothera). This plant is interesting to students of genetics largely for historical reasons. It has been extensively studied both genetically and cytologically, and it was largely on the basis of deVries' studies on this plant that he was led to propose his theory of mutation. For many years Oenothera presented a series of unsolved puzzles —its genetic behavior was atypical and it differed from most

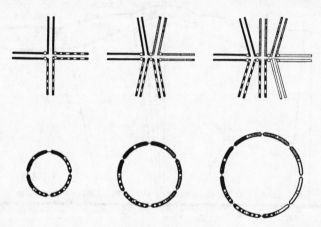

Fig. 79.—Schematic representation of the method of origin of large chromosome rings through successive reciprocal translocations. Pachytene pairing is indicated above. For simplicity all translocation breaks are represented as being immediately adjacent to the centromere. In naturally occurring translocations this is not always the case. As in figure 75, only a single chromatid per chromosome is indicated at diakinesis.

organisms in that its chromosomes were not associated in simple pairs. Finally, as a result of the discovery of reciprocal translocations in Datura by Belling, it was realized that many forms of Oenothera are higher order translocation heterozygotes. The detailed interpretation was worked out by many investigators, including Renner, Cleland, Håkansson, Darlington, and others.

Rings of Many Chromosomes.—An understanding of the basis of inheritance in Oenothera can be gained from a consideration of relatively simple cases in other plants. A simple reciprocal translocation in Datura, in which the two segments involved are approximately whole arms, gives a ring of four chromosomes, which at metaphase is arranged in a figure 8 fashion so that alternate members of the ring of chromosomes go to the same pole. If a second interchange occurs, that involves the second limb of one of the chromosomes of the first translocation and a limb of a third non-homologous chromosome, the pachytene configuration will be a ring of six chromosomes arranged so that alternate members go to the same pole. A third reciprocal interchange can involve another pair of chromosomes, making a ring of eight (Fig. 79). Any number of chromosomes up to the total

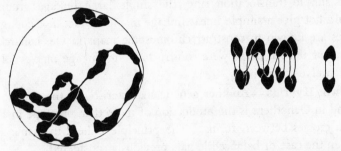

Fig. 80.—Chromosome configurations at diakinesis and metaphase in Oenothera. Here there is a ring of 12 chromosomes plus one pair, such as is found in Oe. lamarckiana and other forms. Note the regular zigzag arrangement at metaphase. (Modified from Cleland and Oehlkers. For clearness the pair at metaphase is moved away from the ring of 12.)

number present in the species (12 pairs in Datura) can theoretically be involved in such a ring.

Complexes.—Such translocations have occurred in Oenothera, resulting in a ring of 12 chromosomes and a single pair in Oe. lamarckiana, the species with which deVries worked (Fig. 80). Because of the zigzag arrangement of chromosomes at metaphase, disjunction occurs in such a way that only two kinds of spores are produced, so far as the chromosomes in the ring are concerned. These two combinations of seven chromosomes (six from the ring and one from the pair) are known technically as *complexes* and are often given names—"gaudens" and "velans" in the case of Oe. lamarckiana. The two homozygous types fail to appear because the velans complex carries one zygotic lethal and gaudens another. Thus, this form is a translocation heterozygote kept heterozygous by balanced lethals.

Linkage in Oenothera.—An important genetic consequence of the balanced condition of Oe. lamarckiana is that there are only two linkage groups. Genes in the free chromosome pair segregate in the normal way, and independently of those in chromosomes involved in the ring of twelve. The genes in the chromosomes in the ring of twelve make up a single linkage group. Crossing over can of course occur in the pachytene arms and exchange segments of chromosomes in opposite complexes. As in the case of a simple translocation ring, the single large linkage groups would not give a simple linear linkage map. In fact, not enough genes are known to construct chromosome maps in Oe. lamarckiana, but if such maps were constructed, the linkage map would be a twelve-armed one.

Twin Hybrids.—Another genetic characteristic of the situation found in Oenothera is the production of more than one type in F_1 from crosses between forms. The principle is here the same as that in the case of balanced lethals previously considered.

Analysis of Complexes.—Other forms of Oenothera have other configurations. There may be a ring of 14, a ring of 8 plus a ring of 6, seven pairs, or any one of the remaining possible combinations of rings and pairs. A systematic study of combinations of complexes from different forms has made it possible to identify, with respect to a standard, the arrangement of segments in many of the complexes. The fact that such studies have given self-consistent results is sufficient evidence of the correctness of the interpretation.

The above account of Oenothera is simplified. There are a number of complications of interest, but only a few can be considered here. It is seen that we have assumed that each chromosome is made up from only two original chromosomes. This will be true only if all the translocations involved result from breaks at the centromere. Actually this is known not to be the situation in all breaks. There is sometimes a segment near the centromere from a third original chromosome. The frequency with which such *interstitial segments* are present is not known. Crossing over between such an interstitial segment and its homolog will result in a partial breakdown of the system, and gives rise to a change sometimes called a *half mutant*. Some of the mutations of deVries were due to crossing over of this kind.

Irregular Chromosome Distribution.—Distribution of chromosomes is not always regular and it is possible to get extra-chromosome types. These usually have phenotypic effects (Chapter XV), and they constitute another portion of the mutants of deVries. Another mutation found by deVries is a dwarf form known as *nanella*. This is due to segregation of a simple recessive gene. The nanella gene is in the gaudens complex, in such a position that there is not often a crossover between it and the translocation point. It is, therefore, in a balanced heterozygous condition. When occasionally it does get into the velans complex by crossing over, the homozygous form appears.

Mutations in Oenothera.—It will be seen that the changes described in Oenothera are, for the most part, not of the type now described as mutations. Though most of them must be referred back to mutations that occurred in earlier generations, the immediate event responsible for the appearance of the new type is a rearrangement of chromosomal material, not the production of a new gene or the loss of an old one. Historically, the term mutation was first based largely on these types in Oenothera. Changes in the significance of technical terms, such as this one, are perhaps unfortunate, but they are inevitable in a rapidly developing subject.

REFERENCES

Anderson, E. G. 1935. Chromosomal interchanges in maize. Genetics, 20:70–83.

Blakeslee, A. F. 1934. New Jimson weeds from old chromosomes. Journ. Heredity, 25:80–108.

Cleland, R. E. 1936. Some aspects of the cyto-genetics of Oenothera. Botan. Review, 2:316–348.

Darlington, C. D. 1937. Recent advances in cytology. 671 pp. Blakiston's Sons and Co., Philadelphia.

Dobzhansky, T. 1936. Induced chromosomal aberrations in animals. pp. 1167–1208, in Duggar, B. M., et al., Biological Effects of Radiation. McGraw-Hill Co., New York. (See also articles by Anderson and Goodspeed in this volume.)

Rhoades, M. M., and B. McClintock. 1935. The cytogenetics of maize. Botan. Review, 1:292–325.

PROBLEMS

1. Diagram the configuration expected following complete conjugation of chromosomes in a reciprocal translocation heterozygote in Drosophila in which the entire right limbs of the second and third chromosomes have been exchanged. Indicate the four arms with the symbols of a few genes known to be located in each one.

2. Draw the chromosome maps that would be expected in a strain of maize homozygous for a reciprocal translocation involving chromosomes 2 and 9 and in which the translocation breaks were at locus 30 in each chromosome. (Maps of only the two altered chromosomes are necessary.)

3. Make diagrams of the chromosomes resulting from reciprocal translocations involving chromosomes 2 and 3 in Drosophila where:
   (a) Breaks in both chromosomes are just to the right of the centromeres.
   (b) Break in 2 is just to the right and that in 3 just to the left of the centromere.
   (c) Break in 2 is just to the left and that in 3 just to the right of the centromere.
   (d) Breaks in both chromosomes are just to the left of the centromeres.

4. Draw the configuration expected with complete conjugation in the translocation heterozygote described in problem 1, in which there is a heterozygous inversion involving the middle third of the right limb of chromosome 2.

5. A Drosophila male heterozygous for a reciprocal translocation in which the right limbs of chromosomes 2 and 3 have been exchanged carries the recessive gene for aristaless (al) in the normal second chromosome and the recessive gene for sepia eye color (se) in the normal third. This male is mated to an aristaless sepia female. What types of offspring are expected? Show by means of diagrams the constitution of all zygotes produced, indicating which ones survive as adults.

6. If a male homozygous for the translocation described above and

heterozygous for both the aristaless and sepia genes (both recessives contributed by one parent of the male) is mated to an aristaless sepia female, what types of offspring will be expected? (Give frequencies.)

7. One of the standard methods of determining which chromosomes are involved in reciprocal translocations in maize depends on the determination of the type of chromosome association at diakinesis of the first meiotic division. If two strains differ by a single translocation, the F_1 between them typically gives a ring of four chromosomes. If they differ by two independent reciprocal translocation (involving different pairs of chromosomes), the F_1 typically shows two rings of four. Given the four known homozygous strains,

> (1) normal
> (2) 1-2 translocation
> (3) 2-3 translocation
> (4) 2-4 translocation
> (X) unknown,

and the following configuration in F_1:

> (1) times (X) — ring of 4
> (2) times (X) — ring of 6
> (3) times (X) — ring of 6
> (4) times (X) — 2 rings of 4

What could you deduce as to the chromosomes involved in the reciprocal translocation in strain X?

8. Starting with a homozygous normal strain of Datura, what is the minimum number of reciprocal translocations required to get a ring of 12 chromosomes at meiosis?

13

CHAPTER XII

MULTIPLE ALLELES

In the preceding chapters we have, in general, dealt with only two alleles at any given locus. This is all that can exist in a single diploid individual, since each locus is then represented only twice. If we consider more than a single individual, it is logically possible to suppose that several different alleles may exist at corresponding loci, in different chromosomes—*i.e.*, that a given + may have several different mutant alleles. In fact, this is frequently the case. The genetic results obtained when *multiple alleles* are concerned may be illustrated by some of the eye-colors of Drosophila. To make this clear, it is convenient first to consider two separate pairs of alleles, *i.e.*, two pairs of non-allelic genes. The wild-type here has dark red eyes. Two eye-color mutants (already mentioned) are white and vermilion, one with pure white eyes, the other with bright vermilion ones. Each of these types breeds true, each differs from wild-type by a single recessive sex-linked gene, and each arose directly from wild-type by a single genetic change (mutation). If the two mutant strains are crossed, the F_1 females are phenotypically wild-type; from which we may deduce that each strain carries the dominant + allele of the recessive mutant gene present in the other. That is, wild-type is $w^+ v^+$, white is $w\ v^+$, vermilion is $w^+ v$, the F_1 is $w\ v^+/w^+ v$. Since both w and v are recessive, the F_1 is then phenotypically wild-type. We have seen that v lies at the locus 33 in the X chromosome; w has been found to lie at 1.5. It follows that, if one breeds from the F_1 female just described, there must be crossing over between the two loci, and $w\ v$ individuals must be produced. Experiments show that such individuals are produced, but are not phenotypically different from $w\ v^+$ —in other words, they have white eyes, and the v gene gives no visible effect when no pigment is present in the eyes (compare the case of black and albino guinea-pigs—Chapter IV).

White Alleles in Drosophila.—This is the kind of result that is to be expected, on the basis of the principles so far discussed; the case is different, however, if we consider the mutant type apricot. This type, with yellowish pink eyes, also arose directly from wild-type, and also acts as a simple sex-linked recessive to wild-type. If the apricot strain is crossed to white and to vermilion, the following phenotypes appear in the F_1 females:

$$\text{apricot} \times \text{vermilion} \rightarrow \text{wild-type}$$
$$\text{apricot} \times \text{white} \rightarrow \text{pale apricot}$$

The second cross gives an intermediate between the two parental types—suggesting at once that neither type has contributed the dominant $+$ allele of the other. If we carry the above crosses another generation, it is found easy to obtain the double recessive apricot vermilion, which is characterized by a light yellow eye-color; but from the cross of white $\times$ apricot no new type can be recovered. It is impossible to get $+$ from a cross of white and apricot, or to produce any double recessive corresponding to white vermilion or apricot vermilion—a double recessive white apricot does not exist.

These relations confirm the hypothesis, based on the phenotype of the F_1, that neither white nor apricot carries the $+$ allele of the other. The simplest interpretation is that apricot is due to a new modification of the w^+ gene—which we may designate w^a. The F_1 female then has the constitution w/w^a; and, since these two genes occupy the same locus, no crossing over is possible between them.

A corollary is evident; if we determine the locus of apricot in the X chromosome it should come out the same as that of white. This is, in fact, the case. While the locus 1.5, was first determined by the use of white, the usual practice now is to use apricot instead, since there results less confusion of phenotypes if other eye-colors are present in the experiment—as shown above for vermilion.

There are a number of other alleles at this same locus—blood, coral, eosin, cherry, tinged, buff, écru, ivory, and several others, some of which are unnamed. All are given symbols with w (from white, the first type studied) as a base: apricot is w^a, eosin is w^e, ivory is w^i, etc. When any two of these types are crossed, the F_1 female is intermediate in eye-color—a relation that is frequent but not invariable for multiple alleles.

Number of Alleles at One Locus.—It may be argued that there are as many distinct alleles as there are distinguishable intermediate grades of color between red and white. This is, however, probably too extreme a statement. But there is evidence that the two extremes themselves each include different types, which are stable and distinct, that can be differentiated by special methods. Muller has shown that certain w^+ alleles differ in the phenotype produced when three alleles are present in the same animal. In general, individuals with three alleles of a given gene can be obtained by making use of duplications, or, as will be seen in later chapters, polyploids. In one case $w^+/w/w$ is red eyed, like w^+/w^+, but a w^+ from a different strain gives (again in $w^+/w/w$ flies) an appreciably paler eye-color. The mutation to white is a relatively frequent one, there being a number of strains phenotypically alike but separate in origin. It has been shown that, if such separate strains be crossed to apricot or to eosin, the resulting w/w^a or w/w^e F_1 females are not always alike; i.e., some whites give more dilution of apricot or eosin than do others. It is likely that similar detailed studies would make it possible to distinguish further members of the intermediate series of alleles.

It must not be supposed, however, that this locus is in a constant state of flux. The changes that occur, while numerous and very diverse, are still relatively rare. Their absolute number is large only because of the very large number of flies examined. In practice, a given homozygous strain breeds true, and needs no special precautions to keep it pure. In experiments, one may safely depend on the constancy of any given allele.

While the case of white is perhaps an extreme one, it is still true that many loci are represented by multiple alleles. The occurrence of more than one mutant allele is the rule in Drosophila, and may be expected in any form where genetic studies are carried out on a very large scale.

Fig. 81.—Self-colored, albino, and Himalayan rabbits. These three types depend on allelic genes.

Albino Series in Rabbits.—An example is that of the albino series in the domestic rabbit. In this form there is a fully pigmented wild-type, a pink-eyed albino with pure white coat, and a type called "Himalayan," with pink eyes and white coat except for the extremities (Fig. 81). Crosses result as follows:

wild-type $\times$ albino $\rightarrow$ F_1, wild-type $\rightarrow$ F_2, 3 wild-type to 1 albino

wild-type $\times$ Himalayan $\rightarrow$ F_1, wild-type $\rightarrow$ F_2, 3 wild-type to 1 Himalayan

Himalayan $\times$ albino $\rightarrow$ F_1, Himalayan $\rightarrow$ F_2, 3 Himalayan to 1 albino

The simplest interpretation is that we are dealing with a set of three alleles—a^+ (wild-type), a^h (Himalayan), and a (albino), with dominance relations in the order listed. All the facts are consistent with this interpretation; but there is a possible alternative. It may be supposed that Himalayan arose first by a mutation from a wild-type gene ($h^+{\to}h$); if now albino arose from a Himalayan strain by mutation of a different wild-type allele ($a^+{\to}a$), its composition would be $a\ h$. The observed results would follow, if one further assumption be made, *i.e.*, that the loci of a and h are completely linked. These assumptions give a formal explanation of the case, and a similar interpretation may be applied in general.

Complete Linkage and Allelism.—In the case of the white eye series of alleles in Drosophila, the hypothesis of completely linked genes meets with difficulties. It is known that white arose from wild-type directly; yet this hypothesis requires that it differ from wild-type by two genes, since it must have the recessive apricot gene present. That is, if we name the two hypothetical mutant genes w and a, wild-type is $w^+\ a^+$, apricot is $w^+\ a$, white is $w\ a$; the fourth type, $w\ a^+$, is unknown. If we add eosin to the series, we must suppose that white represents three simultaneous mutations ($w\ a\ e$, derived directly from $w^+\ a^+\ e^+$), and either eosin or apricot must be supposed to result from the simultaneous mutation of two genes. When the other known members of the series are added to the scheme, it becomes still more complicated. As we shall see in Chapter XIII, mutations may also occur from one mutant allele to another; these mutations necessitate even more complications if one is determined to save the theory of multiple linked genes. On the theory of multiple alleles, however, no such complications arise.

The existence of examples (of which many are known in Drosophila) like that of white makes the interpretation of multiple alleles probable even in such cases as that of the albino and Himalayan rabbits, where the origin is unknown and the number of members in the series is small. The hypothesis of complete linkage is, however, one that must be kept in mind as a possibility.

Strong Linkage.—At the left end of the X chromosome map of Drosophila there is a group of loci that are so close together that they give no crossing over in diploid females—or at least so little that the occasional apparent crossovers produced may be attributed to the occurrence of new mutations. Two of these, scute and achaete, each exist in a series of multiple alleles, and each brings about a reduction in bristle number. They are considered as belonging to two separate allelic series for three reasons: (1) in general the patterns are different, each locus giving mutant types that affect specific bristles; (2) if a recessive scute is crossed to a recessive achaete, the F_1 is wild-type; (3) while the two loci are not certainly separable by crossing over, at least one inversion (known as "scute-8") has one of its points of rearrangement between them. There is, however, at least one mutant ("scute-3") that apparently represents simultaneous mutation in both loci. Because of its existence, the two series were for some years considered to be one; there is a large literature on the nature of this series, based on this natural misapprehension. This case illustrates some of the pitfalls with which the study of multiple alleles is beset.

Nature of Heterozygotes.—The second of the three reasons, enumerated above, for supposing scute and achaete not to be alleles, is of general applicability. We have already used it in introducing the discussion of the white series; white and apricot were presumed to be allelic because they did not give wild-type when crossed. This test is usually adequate, but is subject to several limitations, of which two are particularly important. First, it is desirable to know the origin of the types concerned; in the case of the example cited above from rabbits, the case would be much stronger if we knew definitely that Himalayan and albino had each arisen separately from wild-type. Second, the method is not applicable to dominant mutant genes.

The Black Series in Mice.—The difficulty in the latter case may be illustrated by a case from mice. The wild-type, or agouti, coat color has been referred to before. There exists a recessive mutant type with black coat color, the gene being designated *b*. Another type has a yellow coat, and is due to a gene dominant to the wild-type. These three—yellow, wild-type, and black—behave as though they belonged to a single allelic series (b^y, b^+, b), with dominance in the order named. The argument from the pheno-type of the hybrid between the two mutants (b^y/b) is, however, of no help here, since this hybrid is yellow, as would be expected if two separate but closely linked loci are concerned. The case rests on the absence of crossing over; if a yellow-black hetero-zygote is mated to a homozygous black, no wild-type offspring are produced. There is, however, one further relation that is at least consistent with the view that we are concerned with multiple alleles; the two mutant types differ from the wild-type in the same character—coat color in this instance. This relation is usual for multiple alleles. It may be noted, incidentally, that the hypothesis of multiple allelism was first proposed (by Cuénot) for this particular case in mice.

BLOOD-GROUPS IN MAN

There is an interesting example of multiple allelism in man. The technique of blood transfusion was formerly dangerous, so often leading to death of the patient as to be of questionable clinical value. The studies of Landsteiner and others showed that there are four types of individuals with respect to the nature of the blood; if transfusions are made from a donor who belongs to the same type as the patient, there is no danger of an unfavor-able reaction. This discovery has made transfusion a safe and dependable clinical practice.

Agglutination.—The unfavorable effects of "incompatible" transfusions arise from the fact that the red corpuscles in the blood of one individual may be *agglutinated* (made to gather in clumps) by the serum of another (Fig. 82). The four types are characterized by differences both in the corpuscles and in the serum, as shown by the table on page 201.

The serum carries specific substances, called *agglutinins*. Each different agglutinin reacts with a specific *agglutinogen* in the cor-

Corpuscles from individual of class	Serum from individual of class			
	AB	A	B	O
AB	o	+	+	+
A	o	o	+	+
B	o	+	o	+
O	o	o	o	o

(the + indicates agglutination, o indicates no reaction)

puscles; agglutination results when the corpuscles carry an agglutinogen and the serum contains the corresponding agglutinin. There are two pairs of these; agglutinogens A and B react with

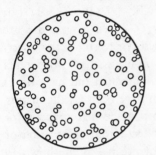 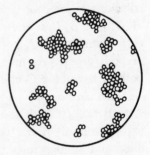

Fig. 82.—Diagrams of human blood corpuscles, magnified. At left, normal blood; at right, agglutinated corpuscles.

agglutinins a and β respectively. The four types of individuals are designated by the agglutinogens (*i.e.*, the types of corpuscles) they possess. The relations then are:

Class	Agglutinogens in corpuscles	Agglutinins in serum
AB	A, B	none
A	A	β
B	B	a
O	none	a, β

Evidently the corresponding agglutinins and agglutinogens are not both present in the same individual—if they were, the corpuscles of the individual would be agglutinated by his own serum. Another relation, which makes the whole scheme easy to remember, is that, if an agglutinogen is absent in the corpuscles of an individual, then the corresponding agglutinin is present in his serum.

Inheritance of Blood Groups.—Analysis of pedigrees of tested individuals has shown that these four classes are dependent on the existence of three allelic genes, of which one conditions the presence of A agglutinogen, one the presence of B, the other the absence of both agglutinogens. The usual system of designating alleles, used in this book, is cumbersome here; it is simpler to designate the three alleles by the letters standing for the agglutinogens they condition—*A, B, O*. Using this system, the constitutions of the four types are: type A B, *A/B;* type A, *A/A* or *A/O;* type B, *B/B* or *B/O;* type O, *O/O*. That is, *A* and *B* are each dominant to *O*, but the *A/B* heterozygote shows the effects of both genes. As may be seen, the result is that two parents both O have only O children; an AB individual can have no O children, and neither of his parents can have been O; the mating A $\times$ B can (if both individuals happen to be heterozygous for O) produce all four classes. It is, in fact, from observation of these and similar rules that the inheritance has been shown to depend on three allelic genes.

The rules for the inheritance of the blood-groups have found an application in medicolegal practice. It is evident that, in cases of disputed paternity, it will sometimes be possible to show that given individuals cannot be related as parent and offspring; the method can never give positive evidence, but in certain cases, easily deduced from the outline given above, it may be concluded that one relationship is possible, another one is not.

The relative frequencies of the four blood-groups differ in different races; this matter will be discussed further in Chapter XVIII, where we shall see that the proportions among the groups furnish confirmation of the view that three alleles are concerned.

SELF-STERILITY

Another example of multiple alleles that has a practical application is that which is concerned in the self-sterility of many plants. The inheritance of this type of self-sterility was first worked out in Nicotiana by East and co-workers; we may take this form as typical of a condition that is widespread among seed-plants. In the particular strains studied, each plant fails to set seed when self-pollinated, though both eggs and pollen are fully fertile when the plant is crossed reciprocally with most other individuals of the population. Genetic analysis shows that this property is due to a series of alleles, known as S^1, S^2, S^3, S^4, etc. Each plant is normally heterozygous for two of these. The rule is that a pollen-tube will not grow well in a style that carries the particular S allele present in the (haploid) pollen grain itself. Thus, in the styles of a plant that is S^1/S^2, neither S^1 nor S^2 pollen grows well, regardless of whether produced by the same plant or by another one; on the other hand, S^3, S^4, etc., grow perfectly. It follows that, if S^1/S^2 is pollinated by S^1/S^3, only the S^3 pollen functions, and there result equal numbers of S^1/S^3 and S^2/S^3 plants. Such tests have been carried out on a large scale, and have given perfectly consistent results. In some species showing this type of self-sterility it is possible to see and measure the growth rate of the two kinds of pollen-tubes. In this way one gets an independent check on the hypothesis; the results are as expected—in pollination of the type $S^1/S^2 \times S^1/S^3$ two distinct growth rates can easily be found in a single style.

Self-sterility is of horticultural importance when it occurs in plants of economic value, and when it occurs in wild populations it presents an interesting special case of the behavior of mixed populations. The physiological study of the relation is likewise a promising field for investigation. These subjects are all being actively studied, but the results do not yet warrant a detailed discussion.

REFERENCES

East, E. M. 1929. Self sterility. Bibliogr. Genetica, 5:331–370.
Morgan, T. H., C. B. Bridges, and A. H. Sturtevant. 1925. The genetics of Drosophila. Bibliogr. Genetica, 2:1–262.

Snyder, L. H. 1929. Blood grouping in relation to clinical and legal medicine. Williams and Wilkins Co., Baltimore.

Stern, C. 1930. Multiple Allelie. Handbuch Vererbungswiss. I, 6. 147 pp.

Timoféeff-Ressovsky, N. W. 1933. Mutations of the gene in different directions. Proc. 6th Internat. Congr. Genetics, I: 308–330.

PROBLEMS

1. If an F_1 female, hybrid between a black and a Himalayan rabbit, is crossed to an albino buck, what types will be expected among the progeny, and in what relative proportions will they be expected? What will an F_1 female, hybrid between an albino and a black rabbit, give among her progeny if mated to a Himalayan buck?

2. If d^t_1 (dwarf) in maize is an allele of d_1, what per cent of crossing over would you expect to get between d^t_1 and cr_1 (crinkly leaf)?

3. Given a stock of Drosophila homozygous for a new eye color mutant that looked like apricot (allele of white), what would you do to test the supposition that it was a white allele? Is it possible to obtain a crucial test in one generation? Explain.

4. The symbol S designates self-sterility alleles in tobacco. What are the progeny expected from the crosses:

$$
\begin{array}{ll}
\text{(a)} & S^1\ S^2\ \times\ S^1\ S^3 \\
\text{(b)} & S^1\ S^2\ \times\ S^2\ S^3 \\
\text{(c)} & S^1\ S^2\ \times\ S^1\ S^2 \\
\text{(d)} & S^1\ S^2\ \times\ S^3\ S^4
\end{array}
$$

5. One of the rules applicable to the inheritance of self-sterility alleles in tobacco states that it is possible for a plant to produce offspring genotypically like itself with respect to S alleles when used in certain crosses as one parent, but not when used as the other parent. From the results of the previous problem, can you state this rule in more specific form?

6. How do you explain the fact that parents both of whom belong to blood group A may have children of group O?

7. What will be the blood groups of the children, where the parents are:

(*a*) O and O	(*f*) A and B
(*b*) O and A	(*g*) A and AB
(*c*) O and B	(*h*) B and B
(*d*) O and AB	(*i*) B and AB
(*e*) A and A	(*j*) AB and AB

8. If you were on a jury before which the case outlined below was being tried, what would you conclude?

Family X claims that baby C, given to them at the hospital, does not belong to them, but to family Y, and that baby D in possession of family Y is really theirs. It is alleged that the two babies, both girls, were accidentally exchanged soon after birth. Family Y denies that such an exchange has been made. Blood group determinations show:

$$
\begin{array}{ll}
\text{X mother} & - \text{ AB} \\
\text{X father} & - \text{ O} \\
\text{Y mother} & - \text{ A} \\
\text{Y father} & - \text{ O} \\
\text{C baby} & - \text{ A} \\
\text{D baby} & - \text{ O}
\end{array}
$$

9. Apricot eye color (w^a) of Drosophila in combination with several other eye colors such as scarlet (st) gives an eye color lighter than either of the single eye colors. Why does this make apricot more useful than white in linkage and other genetic experiments?

CHAPTER XIII

MUTATIONS

THERE are a variety of different changes that may occur in an organism. Assuming that one starts with a diploid form, the possible kinds of new types that it may produce may be classified as follows:

1. Changes in whole sets of chromosomes.
   (*a*) Polyploids.
   (*b*) Haploids.
2. Changes in whole chromosomes.
   (*a*) Trisomics, tetrasomics, etc.
   (*b*) Monosomics, nullosomics.
3. Changes in amounts of portions of chromosomes.
   (*a*) Duplications.
   (*b*) Deficiencies.
4. Changes in relations of parts.
   (*a*) Inversions.
   (*b*) Translocations.
5. Changes in the composition of individual genes.

Any of these changes may be termed mutation, but sometimes only the last—changes in the composition of individual genes—are implied. However, each of the other types may produce a phenotypic effect, and the distinctions indicated here are often difficult to make in practice. The third type and (as will be shown in Chapter XIV) the fourth are especially likely to be confused with true gene mutations. Accordingly, it is desirable to have a term that shall include all possible changes; mutation is used in this sense, with the understanding that changes of whole chromosomes or of whole sets of chromosomes are excluded unless the contrary is specified. In short, the term is made a loose one in order that it shall be possible to use it in discussing actual data.

Frequency of Mutations.—The total number of specimens of Drosophila melanogaster examined by competent observers is at least in the tens of millions; the number of mutations obtained without artificial treatment is in the hundreds or even thousands. Even if these numbers were exact, they would give no adequate measure of the frequency (or rate) of mutation, for different observers differ in their sensitivity to new and unexpected characters, and the likelihood of a new mutation being detected is also related to the system of breeding followed—inbreeding being obviously more likely to bring autosomal recessives to light than is the crossing of remotely related individuals.

A quantitative study of mutation frequency must include some methods of avoiding these two difficulties; *i.e.,* a constant and carefully planned system of breeding must be followed, and some method must be devised for minimizing the differences in personal equation between observers or between different days in the case of a single observer.

LETHAL GENES AND MUTATION FREQUENCY

The most effective method of eliminating the personal equation factor that has yet been developed depends on the determination of the frequency of occurrence of new lethal genes. Crosses are arranged (as will be described below) in such a way that a newly arisen lethal will result in the death of all members of a particular class. The cross is so arranged that all that is needed is the classification of well-known and sharply distinct types. The results obtained in this way are free of the personal equation element, and can be safely taken as showing what any observer would have obtained from the same material.

ClB Method.—The planning of matings that will achieve this result requires some care and ingenuity. There is, however, one standard method, devised by Muller, that is most widely used and that will illustrate the general principles that are present in all efficient schemes for the detection of lethals. This method, available for the detection of lethals in the X chromosome of Drosophila melanogaster, is known as the *"ClB"* method. It depends on the use of a particular type of X chromosome, that has three important peculiarities (see also Chapter X): (1) it carries a long inversion (from about 6.0 to about 59.0 in terms of map units); (2) a lethal is present, as a result of which all males carrying this chromosome die; (3) the dominant mutant gene bar is present, enabling one to distinguish females that carry the *ClB* chromosome, provided the bar gene is not present in the other chromosomes used. The inversion prevents practically all effective crossing over in the X's of heterozygous females, so that the *ClB* combination behaves as a unit in inheritance, as shown in the following scheme:

$$\frac{ClB}{+} \; ♀ \times + \; ♂$$

$$\frac{ClB}{+} \; ♀, \; \frac{+}{+} \; ♀, \; ClB \; ♂ \text{ (dies)}, \; + \; ♂$$

The use of this strain in experiments depends on the fact that each female carrying *ClB* obtained her other X from her father, so that there can have been no lethal in this X in the preceding generation. If a new mutation to a lethal has occurred in this chromosome, the female will now produce no sons, since the *ClB* males always die and the new lethal will cause the other half of the males to die (Fig. 83). The necessary distinction then is, between a sex-ratio of 2:1 (no mutation) and one of 1:0 (new lethal). Mere inspection of the culture is enough for this purpose, with the result that the tests can be carried out on a really large scale.

Frequency of Spontaneous Lethal Mutations.—The use of this test has shown that the frequency of new sex-linked lethals, in untreated material, is somewhat variable but is usually about 2 per 1000 tested chromosomes, which under the conditions of the experiment is the frequency per generation. It is evident at once that this rate cannot apply to all other organisms if it is translated to terms of time. If the ·same frequency per unit time occurred in man, the frequency per generation would average nearly two lethals per X chromosome—and the sex-ratio could

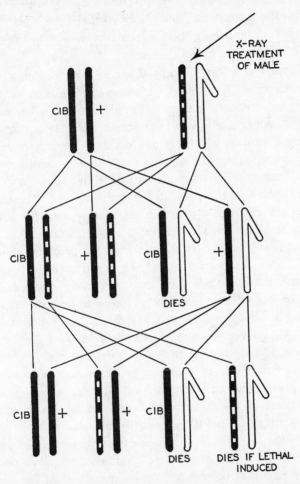

Fig. 83.—Diagram to illustrate the use of the *ClB* method in the study of the induction of new lethal mutations in the X chromosome of Drosophila.

not approach 1:1. It remains possible, though not necessarily true, that the mutation frequency *per generation* is of the same order of magnitude in man and in Drosophila.

x-RAYS AND MUTATION

This rate of production of spontaneous mutations is so low that it is difficult and laborious to study. Muller has shown, however, that this difficulty can be partly overcome by the use of x-rays, which greatly increase the frequency of lethal (and other) mutations. Before describing these effects the physical nature of the effects of x-rays is to be considered.

Mechanism of x-Ray Effect.—When an x-ray photon (gamma ray) is absorbed by an atom, it causes the emission of a fast electron, or secondary beta particle. This electron has a large amount of energy, which it loses in successive steps, each step involving approximately the same loss as does each other one. Each occurs when the fast electron meets an atom and loosens an electron from it. This happens infrequently—about once in a thousand atoms in the path of the fast electron. These latter electrons are endowed with much less energy than the original fast one, and each quickly attaches to a neighboring atom. The atom from which an electron was removed becomes positively charged (since an electron itself has a negative charge), that to which an electron attaches becomes negatively charged. The charged or ionized atoms must be supposed to be responsible for the changes (mutations as well as other effects) produced by x-rays. An ionization results from the release of the original fast electron; but this electron itself induces many secondary ionizations before it loses its energy, so the number of secondaries is much greater than that of primaries. It is, therefore, probable that practically all the effects are due to secondary ionizations.

The intensity of x-radiation is usually measured in terms of the Roentgen unit, or r-unit. This is based on the number of ionizations per cc. of air under standard conditions. A given number of r-units may thus be obtained by intense radiation for a short time, or by weak radiation for a longer time.

The use of the *ClB* technique in x-ray experiments is illustrated in figure 83. A rayed male is mated to an untreated *ClB* female, and the *ClB* daughters are mated individually to normal males. Each such daughter of the treated male constitutes a test of one treated X. If a new lethal is induced by the treatment, then the daughter that gets the mutated chromosome will produce no sons.

Dosage and Mutation Frequency.—Many investigators have studied the relation between the amount of *x*-ray dosage given to males and the frequency with which their *ClB* daughters carry new lethals in treated X's. It turns out that the relation is (for

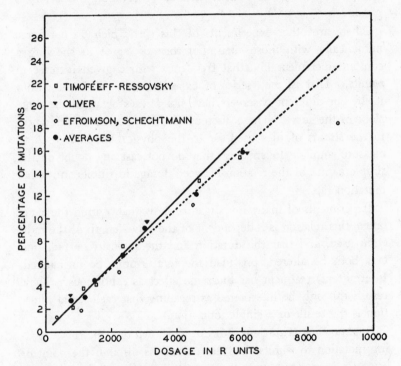

Fig. 84.—The relation between *x*-ray dosage and frequency of induced lethal mutations in the X chromosome of Drosophila. (From Timoféeff, Zimmer, and Delbrück.)

low dosages) a linear one; on the average each increment of 350 *r*-units in the dosage adds 1 per cent to the number of recovered lethals. In the diagram shown in figure 84, this linear relation is shown by the solid straight line. The data evidently fit for low dosages, but give fewer recovered lethals at higher dosages. This was to have been expected, since the method used does not distinguish between chromosomes with one new lethal and those in which two or more new lethals have been induced. As the total number of lethals increases there will, of course, come to be more and more such cases of more than one lethal in a single X. The dotted curve shown in the figure is calculated on this basis, assuming that lethals are distributed among the chromosomes at random; evidently the agreement between observation and theory is good.

There are other experiments of this sort which do not agree numerically with those cited; but they do agree in showing a constant increment in lethal frequency with a given increase in *r*-units. That is, each series of experiments is consistent within itself, but different observers have not always obtained the same slope of the curve. These discrepancies may be due to differences in the strains of flies used, or in the physical methods used to measure *r*-units. In any case they do not cast any doubt on the simple nature of the relation between dosage in *r*-units and lethal mutation rate.

Experiments of many investigators show that, within the *x*-ray range, this relation is independent of the wave-length of the radiation used, and that the duration and the intensity of radiation may both be altered, provided the total *r*-units be unchanged. Intermittent treatment has the same effect as continuous. These results can only be interpreted as meaning that each lethal mutation is the result of a single ionization.

Spontaneous Mutations.—It can be shown that the curve relating mutation to *r*-units does not quite pass through the point of origin; *i.e.*, at zero *r*-units there are still some mutations present. In fact, the natural ionization present under normal conditions cannot account for any large fraction of the "spontaneous" mutations referred to above—about 2 per 1000 X chromosomes per generation.

Induced Lethals and Chromosome Rearrangements.—Lethals are the most easily studied mutant types in this kind of investigation, both because they are relatively frequent and because their detection can be reduced to a routine "fool proof" technique.

There are, however, some possible objections to the application, to other mutations, of results obtained from the study of lethals. We have seen that lethals may be due to losses of materials—to deficiencies. Since x-rays are known to increase the frequency of chromosome rearrangements, it is pertinent to inquire whether, in studying frequencies of lethals, we are dealing with frequencies of short deficiencies or of gene mutations. Study of the salivary gland chromosomes of x-ray induced lethals shows, in fact, that many of them are due to short deficiencies (Demerec). The actual proportion due to gene mutations can scarcely be estimated —it is possible that this proportion is very small.

These considerations raise the question: do x-rays actually induce changes in individual genes at all, or are they effective only in causing breakages and rearrangements of parts of chromosomes, thereby indirectly bringing about inherited phenotypic effects without changing genes? This question is difficult to answer. Direct cytological study of the chromosomes of induced mutations often fails to show any detectable change; but even the salivary gland chromosomes do not permit a sufficiently exact analysis to make such negative evidence altogether conclusive.

Induced Non-lethal Mutations.—Another method of attacking the problems in this field is through the study of induced mutations that are not lethal. Muller has pointed out that mutations may be classified according to a system analogous to that which is used for light-rays. At one end of the "spectrum" there are mutant types that have only slight effects, detectable only by special techniques—presumably others that cannot be detected at all. These slight types include the modifiers that we shall discuss in connection with selection; they may be considered as analogous to the infra-red region of the light spectrum. Next come the "visible" mutant types, not sharply distinct from the slight ones. This class includes all those mutant types that are viable and are definitely distinguishable from wild-type. Analogous to the ultra-violet region of the spectrum are the lethal mutations. This rough analogy fits the case in another respect: there are more slight and more lethal mutations than there are "visibles," just as the visible range occupies only a small portion of the whole spectrum. It should be emphasized that this is a classification of mutant phenotypes, and has no necessary relation to the amount or kind of change in the genes themselves.

Slight Mutations.—Timoféeff and Muller have each utilized the *ClB* technique in a study of the slight mutations. They have shown that slight, but permanent, changes in viability (mostly decreases, only rarely increases) are produced by *x*-rays. Timoféeff estimates that such slight changes are from 2 to 2½ times as numerous as are new lethals. Presumably these are only a fraction of the "slight" types that are produced, the remainder very likely having no detectable effect on viability.

Frequencies of Lethal and Non-Lethal Mutations.—The mutations resulting in visibly distinct phenotypes are difficult to estimate quantitatively, since the personal equation element cannot be entirely eliminated here. Their detection can be made more nearly objective by the use of the *ClB* technique. The scheme shown in figure 83 shows that all the sons of a tested daughter of a rayed male will carry a given rayed X; if a mutation was induced in this X, all these sons show the new character. In this case the personal equation is reduced to a minimum, since what is required is to detect a character present in many specimens of a family, rather than in a single individual.

Direct comparison of lethals and "visibles" obtained in the same series of experiments indicates that there are about 9 times as many lethals as phenotypically distinct types; this proportion must not be taken as exact, but only as giving a rough idea of the relative frequencies.

Attached-X Method for Sex-linked Mutations.—There is another method for the detection of sex-linked "visibles." If a treated male is mated to an attached-X female, each son gets a single treated X, and will show immediately any new sex-linked viable and distinguishable character.[1] This method is more ef-

[1] No special technique is required for the detection of newly arisen dominant genes, since they produce their characteristic phenotypic effects at once, and it is not necessary to control the constitution of the homologous locus. The frequency of dominants is, however, so low that they play little part in the study of mutation.

ficient, since it does not require the use of a separate culture for the testing of each treated chromosome. Its disadvantages are that it does not minimize the personal equation factor, and that it cannot be used for the study of lethals. Lethals are evidently to be expected; they will cause the death of some males, thereby modifying the 1:1 sex-ratio. The frequency of lethals, however, is in general too low to be distinguished from the effects of sampling on the ratio. Furthermore, any lethals induced in the X are at once lost, and cannot be tested further, as can those detected by the *ClB* method.

Induced and Spontaneous Mutations.—One question that has been studied by means of the "visible" types is: do *x*-rays induce the same mutant types as those that arise spontaneously? The answer here is evidently, yes. It remains uncertain if the relative frequencies of the various possible mutations are the same under the two conditions; but there can be no doubt that most of the types induced by *x*-rays may also occur spontaneously, and that most of those that occur spontaneously may be induced. The results obtained by Stadler with maize and by various investigators with Drosophila make it most probable that the relative frequencies for different genes do differ between irradiated and untreated series; but it is not possible to estimate accurately the degree of the differences present.

Induced Reverse Mutation.—As pointed out above, it has been suggested that *x*-rays do not induce gene-mutations, but only cause breakage and reunion of chromosomes. Several investigators (*e.g.*, Muller, Patterson, Timoféeff) have used the "visible" types in an attempt to solve this problem. These studies have shown that *x*-rays may induce reverse mutations; that both $a^+\rightarrow a$ and $a\rightarrow a^+$ may be brought about. Specifically, $f^+\rightarrow f$ (forked) has been induced, and then the induced f has been rayed and has produced f^+. This result leaves little doubt that actual gene changes are concerned.

Mutations at the White Locus.—The most elaborate series of experiments of this type are those of Timoféeff with the alleles of white. Some of the kinds of changes that he has induced are

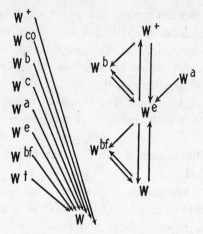

Fig. 85.—Some of the kinds of mutations that have been found at the white locus in Drosophila. (After Timoféeff.)

shown in the diagrams (Fig. 85). Using a dosage of 4,800 *r*-units the following frequencies were obtained (w^x signifies any new alleles):

$$w^+ \rightarrow w, \quad 5.1 \text{ per } 10,000$$
$$w^+ \rightarrow w^x, \quad 7.6 \text{ per } 10,000$$
$$w \rightarrow w^x, \quad 0.5 \text{ per } 10,000$$

Adding these and other results, we can summarize the relations as follows:

all decreases in color intensity, 4.8 per 10,000
all increases in color intensity, 0.4 per 10,000

These results show that different genes, even different alleles at a single locus, may have quite different mutation rates. These frequencies are, if anything, higher than those for most genes. It must be remembered that they were obtained by the use of a dosage that would have raised the rate for sex-linked lethals from the spontaneous value of 2 per 1,000 to about 135 per 1,000. The frequencies for most individual genes are so low, without treatment, that their study cannot be carried out. There are, however, a few "multimutating" genes known, that give very high frequencies of mutation. These may represent special cases; they will be referred to again in the next chapter. It may be noted here that, in the ones studied by Demerec, their mutation frequency has not been increased by x-rays.

Other Factors Influencing Mutations.—Many methods have been employed in attempts to increase mutation frequency. No other as effective as x-rays (and the equivalent gamma-rays of radioactive elements) has been found. There can be no doubt that ultra-violet light and increased temperature both increase mutation frequency. The first is technically difficult to use, since tissues are relatively opaque to it, with the result that it is difficult to get it into the germ-cells; high temperature increases the mutation frequency so little (not over a doubling per generation per 10° C. increase in temperature[1]) that the study of the effect is extremely laborious. Chemical methods may be effective, but the results reported do not yet seem conclusive.

It is probable that the effectiveness of x-rays depends on the fact that they produce large disturbances within very small regions. Other methods of treatment, such as the use of extreme high temperatures or poisons, have much more general effects; if they produce effects comparable in violence with those due to x-rays, in the neighborhood of a given gene, they will commonly produce them over a large area, and death of the cell (if not of the individual) will result.

REFERENCES

Duggar, B. M., et al. 1936. Biological effects of radiation. 2 vols., 1,343 pp. McGraw-Hill Co., New York.

Oliver, C. P. 1934. Radiation genetics. Quart. Review Biol., 9: 381–408.

Stubbe, H. 1938. Genmutation. Handbuch Vererbungswiss. 2, F. 429 pp.

Timoféeff-Ressovsky, N. W. 1934. The experimental production of mutation. Biol. Reviews, 9:411–457.

Timoféeff-Ressovsky, N. W. 1937. Experimentelle Mutationsforschung in der Vererbungslehre. T. Steinkopf. Dresden and Leipzig. 181 pp.

[1] This is the increase per generation. Since higher temperatures also shorten the development period, the increase per unit of time is greater.

PROBLEMS

1. Methods for detecting sex-linked lethal and sex-linked "visible" mutations and measuring their frequencies were described for Drosophila. Outline a technique for measuring the frequency with which dominant mutations occur. How would you distinguish between sex-linked and autosomal dominants?

2. Outline the method you would use for determining the locus of a new dominant mutation in Drosophila. Do the same for the locus of a new recessive mutant.

3. Outline the method of determining the frequency of reverse mutations of the recessive eye-color gene scarlet (st) in Drosophila with a specified x-ray treatment of mature sperm, $i.e.$, the frequency of mutation of st to $st+$.

4. If dominant lethal mutations are produced with a relatively high frequency following treatment of mature Drosophila sperm with x-radiation, what would be the effect on the sex ratio when treated males are mated to attached-X females? What would be the effect on the sex ratio when such males are mated to normal females?

5. x-Ray treatment induces mutations at random; that is, these rays cannot be "directed" so as to produce specific mutations. On this basis, how would you set out to obtain new mutant alleles of the eye color gene light (lt) in Drosophila by using x-ray treatment? If in getting these you make use of an already existing allele of lt, devise a way of differentiating the old and the new alleles without assuming their phenotypic effect to be different.

6. The miniature (m) and dusky (dy) characters in Drosophila are phenotypically similar and the genes that differentiate them lie close together in the X chromosome. A female heterozygous for both miniature and dusky is wild-type, suggesting that the m and dy genes are not allelic. What other type of evidence could be used to show that this is true? In outlining an experiment for doing this, remember the possibility of mutation. If it is necessary to do so, make use of genes relatively closely linked with miniature and dusky.

7. If mature pollen of maize homozygous for the normal allele of the waxy gene is x-rayed and then used in crosses with homozygous waxy plants, occasional kernels will be produced that are phenotypically waxy. Presumably this is due to mutation and the endosperms of the waxy kernels are therefore $wx/wx/wx$ instead of $+/wx/wx$. If such kernels are planted it will be found

that most plants that come from them are $+/wx$ and not wx/wx, as the phenotypes of the endosperm might be taken to indicate. How would you account for this apparent discrepancy?

8. Aside from the fact that long exposure to x-rays may result in serious "x-ray burns," why should persons who habitually work with x-ray equipment be careful to shield themselves from the rays?

9. In barley grains the embryo is differentiated into stem and root, but usually each of the separate heads of a barley plant is developed from a single cell present in the seed. If ungerminated seeds are x-rayed, mutations are produced. Knowing that barley is normally self-fertilized, how would you go about detecting induced recessive mutations? To make this simpler, include only mutations that affect chlorophyll and are therefore detectable in the seedling stage.

CHAPTER XIV

POSITION EFFECT

BAR IN DROSOPHILA

IT was pointed out in Chapter XIII that there are a few genes known that have high mutation rates, and it was suggested that possibly these represent special cases, not directly comparable to ordinary mutation. One such case has been analyzed, and has in fact been found to represent a special case. The analysis, now to be presented, has led also to certain inferences concerning the

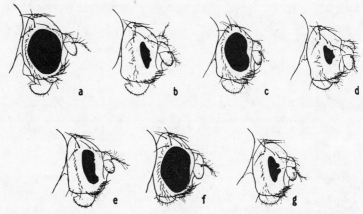

Fig. 86.—Eyes of Drosophila melanogaster. a, wild-type female; b, homozygous bar female; c, heterozygous bar female; d, double-bar male; e, homozygous infrabar female; f, heterozygous infrabar female; g, double-infrabar male.

manner of action of genes, and has raised new problems concerning the nature of ordinary mutations.

Bar Reversions.—The bar type of Drosophila has already been referred to. This form was discovered by Tice (1914). It was shown by May and Zeleny that homozygous strains of bar are not quite constant; about 1 in 1,600 offspring from such a strain carry the B^+ allele, and a similar proportion carry a new allele, called double-bar, that conditions an eye smaller than that of bar (Fig. 86). Sturtevant and Morgan showed that the occurrence of these new types is related to crossing over; this relation will be elaborated below.

Cytological Characteristics of Bar Types.—Muller and Bridges have studied the salivary gland chromosomes of the three types, bar, double-bar, reverted bar, with the results indicated in figure 87. Bar has a section of about four bands, which is present in wild-type, represented twice. This "repeat" is adjacent to the normally present section, and the bands are arranged in the same

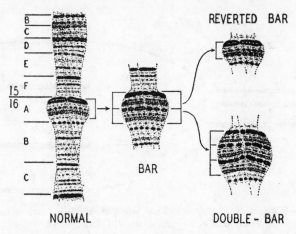

Fig. 87.—The bar region of X chromosomes from salivary glands, showing the duplicated section present in bar. (From Bridges.)

order in the two sections. Double-bar has this same section represented three times; reverted bar is indistinguishable from wild-type in this respect, as it is in every other that has been studied. Bar may thus be considered as due to a special type of duplication; double-bar is due to the presence of this same duplication in two doses. The genetic results, now to be discussed, had already indicated part of these results—they had, in fact, led to the application of the name double-bar. They also show the mechanism by which the changes (except that from wild-type to bar) occur.

Unequal Crossing Over.—The recurrent changes at this locus occur only in females, and can be shown to be associated with crossing over. For this purpose, use is made of the two mutant genes forked (*f*—locus 56.8) and fused (*fu*—locus 59.5), lying one on each side of bar, and close to it. If females of the constitution *B/f B fu* are tested, it is found that all the reversions (*B+*—*i.e.*, with no duplicating segment) and all the double-bars (*BB*—*i.e.*, with two duplicating segments) are either *f fu+* or *f+ fu*—*i.e.*, they are crossovers in the *f-fu* interval that includes the bar locus, though such crossovers constitute less than 3 per cent of all the offspring. These results are due to *unequal cross-*

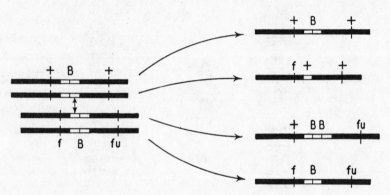

Fig. 88.—Unequal crossing over and the origin of double-bar and reverted bar. Position of crossover indicated by double arrow.

ing over, a phenomenon so far known only in this case. As figure 88 indicates, this is probably to be attributed to conjugation between the left section of one bar section and the right section of its homolog.

Other changes occur here, likewise together with crossing over between the loci of *f* and *fu*: *BB/B→B+*, *BB/BB→B* and *B+*, *BB/B+→B*. These changes may be interpreted by means of diagrams of the same nature as the one shown. It is evident that *BB/B* should give rise to triple-bar, this being the complement of the *B+* that is actually recovered; likewise *BB/BB* should give rise to quadruple bar. The former has not been obtained, presumably because it is relatively inviable; but Rapoport has shown that quadruple-bar (*i.e.,* four duplicating sections) occasionally survives. The specimens obtained were weak and sterile; their eyes had from 7 to 10 facets each, compared to about 26 in the double-bars of the same strain.

Infrabar.—Still another change has been found that gives a less extreme small eye known as infrabar. This change, which took place in a chromosome that was already *B,* was probably a true gene mutation. It has occurred only once; the salivary gland chromosomes have not been fully studied, but are probably not different from those of bar. Infrabar behaves like bar; it gives reversions to wild-type, and it gives double-infrabar, but bar cannot be recovered from it. From a female that carries both bar and infrabar (B/B^i) it is possible to obtain (again by crossing over between f and fu) two new double types—BB^i and B^iB (bar-infrabar and infrabar-bar). These are phenotypically indistinguishable, and have eyes only a little larger than those of double-bar. Genetically they may be distinguished by the usual test for sequence of genes, as follows: BB^i/f fu gives f B^i and B fu, whereas B^iB/f fu gives f B and B^i fu.

Before the genetic and cytological proof of unequal crossing over resulting in duplicating segments, the case of bar was a typical example of frequently recurring mutation. As will be seen, it is now clear that the mechanism involved is one that changes quantities, not qualities, of genes. It is probable that the numerous kinds of changes listed above include only one gene mutation in the strict sense (that from B to B^i).

Other Frequently Mutating Genes.—The question is therefore raised—do some or all other examples of frequently mutating genes represent some similar type of quantitative change? It is clear that none of the adequately studied cases can be due to exactly the same mechanism as that which is acting in bar; either the changes occur in somatic tissue, or they can be shown to occur without crossing over in the immediate neighborhood of the locus concerned. Nevertheless, there remains a strong suspicion that similar kinds of changes in gene quantities may be responsible in many such cases.

Rhoades has described a case in maize in which such an interpretation is unlikely. One of the most widely studied recessive
genes in maize is *a*, which is concerned in the production of
aleurone color; *a* plants have colorless aleurone, a^+ plants have
colored aleurone if certain other specific genes (at other loci) are
also present. The recessive *a* is ordinarily quite constant; but
Rhoades has studied a strain in which mutations from *a* to a^+
occur with a high frequency, several in each kernel, so that the
kernels show dots of color (cells with a^+) on a colorless (*a*)
ground. This high frequency is due to the presence of another
gene, called *Dt,* that is not even in the same chromosome as *a*.
No other phenotypic effect of *Dt* is known, but it consistently
transforms *a* from a constant gene to a highly mutable one. No
chromosome differences have been detected between any of the
plants here concerned and normal maize plants. It is difficult to
avoid the conclusion that the changes here are true gene mutations. The question of how many "multimutating" genes really
undergo gene mutation in the strict sense remains an open one.

POSITION EFFECT

The bar case has led to another result that is of importance in
the study of mutation and of development. The eyes of Drosophila are made up of separate ommatidia, each of which is
visible at the surface as a single facet. These facets are relatively
constant in size and their number therefore forms a convenient
measure of the size of an eye.

Facet Number.—In the case of the bar series of eyes, it has
been shown that both temperature and a series of minor genes at
other loci affect the number of facets. If these two variables are
made reasonably constant, it is possible to compare the effects of
the various bar alleles on facet number. Under the conditions
actually encountered in one such series, the wild-type flies had
about 750 facets (females 779, males 734), *B/B* females 68, *B/+*
females 358, *BB/BB* females 25 (the values given are approximate averages).

It would be expected that *B/B* would have the same number

of facets as $BB/+$, since each has two duplicating sections. This expectation is not realized; $BB/+$ regularly has fewer facets than B/B. This is to say, two duplicating sections in one chromosome are more effective in decreasing facet number than are the same two when they lie in separate homologous chromosomes. There are three available comparisons, all showing this same relation:

$$BB/+ \; = \; 45.4 \qquad B/B \; = \; 68.1$$
$$BB^1/+ \; = \; 50.5 \qquad B/B^1 \; = \; 73.5$$
$$B^1B^1/+ \; = \; 200.2 \qquad B^1/B^1 \; = \; 292.6$$

This phenomenon, known as the *position effect*, constitutes a demonstration that the effectiveness of a gene may be a function, not only of its own constitution, but also of the position it occupies with respect to neighboring genes.

Position Effect and Hairy.—Another example of the same relation has been recorded by Dubinin and Sidorov. A translocation involved a break in the third chromosome near the locus of the recessive hairy (h—extra hairs on wings and elsewhere). Tests showed that the translocation gave the phenotypic effects to be expected if it contained the h gene (*i.e.*, translocation/h showed the hairy character). Dubinin and Sidorov were able to get crossing over between the locus of h and the translocation break. They showed that, if the allele present at the hairy locus in the translocation chromosome was transferred to a normal chromosome, it now acted in all respects like a h^+ gene; whereas a h^+ allele from a normal chromosome, when put into a translocation chromosome, gave the same hairy phenotype in translocation/h. Here, as in the cases of bar and infrabar, there is no permanent change in the gene, which can be recovered in its original form when it is put into its original position. A case similar in every way to that of hairy has been reported for curled (wing character, third chromosome) by Panshin.

Mechanism of Position Effect.—In the above cases there seems no escape from the conclusion that we are concerned, not with changes in the genes, but with modifications of their effects on development. A possible model, to illustrate how such modifications may come about, may be based on diffusion. If two genes each produce a specific substance, and each of these substances diffuses out from the gene that produces it, we may suppose that each is rapidly destroyed in the cell. There will result concentration gradients, with most of each substance near its point of origin. If the two substances interact, the amount of their joint product will depend on the distance that lies between the two genes concerned. If the joint product influences the course of development, we have all that is needed to picture a possible mechanism for the position effect.

Other Possible Cases.—There are other cases that suggest the existence of position effects. In none of them has final proof been furnished by the recovery of the gene concerned and the demonstration that its new properties are due to its new position. For example, a high proportion of the translocations found in Drosophila are inviable when homozygous. It is possible that this lethal effect is due to loss or mutation of some gene or genes lying near the breakage point; it is also possible that there is still present a complete set of unmutated genes, but that some of them have position effects, due to their new locations, that are fatal to the organism.

Position Effect and Mutation.—In view of the existence of the three established cases (those of bar, hairy, and curled) it is probable that many of the other suspected examples do in fact represent position effects. This conclusion is of importance in connection with the general theory of mutation. It has been suggested that perhaps all mutations (or all x-ray induced mutations) are due to losses or gains of genic materials. The position effect indicates another possibility: it has been suggested that perhaps all mutations (or, again, only all x-ray induced mutations) are due to position effects. We are not inclined to take this hypothesis seriously, but it must be admitted that, in any given case, it is practically impossible to assert confidently that one is dealing with a change in the composition of a gene. It can often be established by cytological observation that the amount and arrangement of material has undergone no detectable change when a mutation has occurred; but the difficulty is with the one word "detectable." The available methods are not adequate to exclude the possibility of minute rearrangements; in fact rearrangements of the minimum extent one could hope to detect have been found. Thus, Bridges has shown that the dominant mutant type dichaete is associated with an inversion that includes only three bands of the salivary gland chromosomes.

Much remains to be done in the study of the position effect, which is at present one of the most rapidly developing branches of genetics.

REFERENCES

Demerec, M. 1935. Unstable genes. Botan. Review, 1:233–248.
Dobzhansky, T. 1936. Position effects on genes. Biol. Reviews, 11: 364–384.

PROBLEMS

1. From the information given in the text about the dotted aleurone gene (Dt) in maize, what ratio would be expected among the F_2 kernels from the cross of $a + \female \times + Dt \male$?

2. The gene a in maize differentiates brown plants and anthers from purple plants and anthers (if the dominant alleles of the B and Pl genes are present) as well as colorless from colored aleurone. In the presence of Dt, a mutates to a^+ in sporophytic tissue as well as in endosperm tissue. Assuming the rate in the sporophyte to be of the same order as that in the endosperm, what would you expect to be the appearance of an $a\ Dt$ plant? Rhoades has shown that an a^+ gene that arose from an a gene in the presence of Dt, is stable. What methods could have been used in obtaining such an a^+ allele?

3. From a $\dfrac{f\ B\ +}{+\ B^i\ fu}$ Drosophila female a forked fused male with narrow bar eyes is obtained. Is it $f\ B\ B^i\ fu$ or $f\ B^i\ B\ fu$?

CHAPTER XV

NON-DISJUNCTION AND RELATED PHENOMENA

IT is the general rule that the two members of a pair of homologous chromosomes are distributed to separate haploid cells at the meiotic divisions, but exceptions occasionally occur. It sometimes happens that both members of a particular pair of chromosomes are included in one first division daughter nucleus, leaving no representative of this pair in the sister nucleus. This failure of chromosomes to disjoin properly is known as *non-disjunction*. At the time this term was first used, the cytological basis of such irregularities in meiosis was not well understood, and it has turned out that the term is misleading. Actually, the irregular distribution is almost always the result of failure of metaphase pairing rather than failure of disjunction of a normally associated pair of homologs. The failure of metaphase pairing may result from failure of conjugation at zygotene, or from complete separation at diakinesis. The two unpaired homologs— *univalents*—are distributed to the two poles more or less at random with respect to each other. They may go to opposite poles or to the same pole. The term *non-conjunction* has been proposed for this failure of pairing and subsequent irregular distribution, but the term non-disjunction is so widely used in the literature that we shall continue to use it with the understanding that its meaning is extended to include non-conjunction.

Frequency of Non-disjunction.—The frequency of non-disjunction varies in different forms and for different chromosomes in a given strain. In plants where metaphase pairing is by chiasmata, there is often a rough relation between shortness of chromosomes and frequency of non-pairing at metaphase. There may also be variations in frequency of non-disjunction among different individuals of the same species. Genes that influence the amount of metaphase pairing of chromosomes are known in a number of forms, and it is known that environmental factors such as temperature influence such associations. Under normal conditions the frequency of irregular distribution for a given pair of chromosomes is not higher than one irregularity in several thousand divisions. Considering its complexities, the meiotic mechanism is remarkably stable.

TRISOMIC TYPES

As a result of non-disjunction at meiosis, gametes or spores with an extra chromosome may be formed. If such gametes or spores function, individuals will be produced with the diploid number of chromosomes plus one extra whole chromosome. Such an individual may be called a *trisomic* type or it may be referred to as a "triplo" type, specifically designated by adding the name of the particular chromosome represented three times—for example, triplo-X or triplo-4 in Drosophila, triplo-10 in maize, or triplo-F in tobacco. In still other cases, a trisomic type may be referred to as a 2n plus 1 (2n + 1) individual, n designating a haploid chromosome set.

Trisomics in Plants.—Trisomic types are known in a number of plants; because their meiotic behavior is better understood here than in animals, we shall consider them first. With a particular chromosome represented three times instead of two, conjugation becomes more complicated. In maize and several other plants, conjugation, under these conditions, is by twos just as it is with only two homologs present. The essential difference is that at any given point, one homolog is unpaired. Usually pairing partners change one or more times so that, considering the entire length of the chromosomes, each one of the three homologs is paired over some part of its length (Fig. 89). In other words, conjugation occurs as though pairing affinities were completely saturated as soon as two chromosome threads pair. A group of three homologous chromosomes associated during the first division of meiosis is known as a *trivalent*. Considering the entire trivalent, crossing over can occur between chromatids of any two homologs. But at one level chromatids from homologs 1 and 2 may cross over, whereas at another level, those from 2 and 3 or 1 and 3 may undergo crossing over (Fig. 89). Correspondingly, chiasmata may be formed between any two homologs that happen to be conjugated at a particular level.

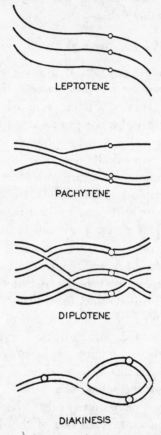

Fig. 89.—Meiotic prophase behavior of three homologous chromosomes in a trisomic individual.

Distribution of Chromosomes of a Trivalent.—Separation of chromatids by pairs at diplotene gives rise to trivalents, the forms of which depend on the number and position of chiasmata. As in bivalents, terminalization varies in different species. In Datura, in which terminalization is complete, trivalents of such forms as those shown in figure 90 can be seen at diakinesis. At metaphase the trivalent is arranged so that two of the three homologs are oriented toward one pole and one toward the other. Of the daughter nuclei one has an extra chromosome and the other has a normal haploid set. Finally, two of the four nuclei resulting from the two meiotic divisions carry the extra chromosome and two are normal. With distribution of this type, the extra chromosome would be expected to be transmitted to half the offspring in the next generation.

Extra Chromosomes and the Phenotype.—Considering the genetic consequences of the trisomic condition, it is found first of all, that there is usually a characteristic modification of the phenotype of a trisomic individual. Thus, in Datura strains in which no mutant genes are present, trisomic plants can be identified

Fig. 90.—Diakinesis trivalent configurations of the "rolled" trisomic of Datura. (From Belling.)

phenotypically. Furthermore, it is found that the phenotypic effects are characteristic for each one of the twelve different chromosomes. A plant trisomic for chromosome 1 differs from the wild-type and from the other eleven possible trisomic types, all of which are known. In some plants, maize for example, not all of the trisomic types are easily identifiable by their phenotypic appearances. This phenotypic modification of a plant carrying an extra whole chromosome has been shown to be the result of a change in *genic balance*. This means that, although all the genes are present, they are not present in the normal proportions. Those in one chromosome are present three times while those in other chromosomes are represented only twice. The result of such a disturbance in the balance depends on two intrinsic factors: (1) the relative proportion of genes that are represented three times and (2) the particular genes so represented. These two factors explain why each trisomic type deviates from normal in a particular way. Other examples of the influence of disturbances in the normal genic balance have been seen in duplications, in deficiencies, and in combinations of the two. Still other examples will be considered later, and we shall return to further considerations of the balance concept.

Transmission of Extra Chromosomes.—In many cases in plants extra chromosomes are not transmitted to the following generation with the frequency expected with the typical segregation described above. As an example, an extra chromosome 10 in maize is transmitted to about a third of the offspring when the trisomic plants are used as female parents in crosses with normal diploids. In the reciprocal cross, 2n × 2n + 1, only about one per cent of the offspring carry the extra chromosome. The low frequency of transmission in the latter case is largely the result of pollen competition—n + 1 grains are potentially functional, but are not successful in competition with normal haploid grains. This again is a manifestation of the disturbance in the normal balance of genes—in this instance in the gametophyte generation. The low transmission through the female gametophyte may be the result of frequent failure of one of the three like homologs to be included in a daughter nucleus at the first meiotic division, or it may be the result of competition among the four megaspores. Some lagging of chromosomes does occur at anaphase of the first division because of failure of trivalent formation in a proportion of sporocytes. With a bivalent and a univalent instead of a trivalent, the univalent is not always included in a daughter nucleus, but may remain in the cytoplasm and disintegrate.

Trisomic Ratios.—With three homologs instead of two, all loci in the trisomic chromosome will be represented three times. With two different alleles of a given gene, four genotypes, instead of the usual three, are possible, viz., $+/+/+$, $+/+/a$, $+/a/a$, and $a/a/a$. Obviously, with three alleles segregating, genetic ratios will be different from those in normal diploids. Random distribution of the three alleles $+/+/a$, with no crossing over between the locus of *a* and the centromere, will give the following spores or gametes: $1 + + ; 2 + a ; 2 + ; 1 a$ (Fig. 91). With dominance of one normal allele over two recessives, a testcross to a recessive diploid *a/a* will give a phenotypic ratio of 5:1 for $+$

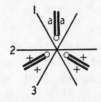

PLANE	METAPHASE ARRANGEMENT	FIRST DIVISION PRODUCTS	SECOND DIVISION PRODUCTS

Fig. 91.—Types of segregation in a trivalent, one chromosome of which carries a mutant gene, "a."

and *a* respectively. Such a ratio is seldom obtained in plants because of the fact that the extra chromosome is not transmitted to half the offspring. The observed ratio will depend on just how frequently the extra chromosome is transmitted and on the frequency of crossing over between the locus concerned and the centromere. For the tenth chromosome trisome in maize, McClintock and Hill obtained a ratio of 3.8 R to 1 $+$ (colored and colorless aleurone) from the cross $R/R/+$ ♀ $\times$ $+/+$ ♂. This ratio is consistent with the direct observation that the extra chromosome is transmitted to about 33 per cent of the offspring. The reciprocal cross, $+/+$ ♀ $\times$ $R/R/+$ ♂, gives a ratio of 2 colored to 1 colorless. An actual cross of this kind gave 646 R to 355 $+$ kernels. The basis of this ratio becomes clear when we write down the constitutions of the spores expected (2 R; 1 $+$; 2 $R/+$; 1 R/R, disregarding crossing over between the R locus and the centromere), and remember that ordinarily only those with a single tenth chromosome function. The cross $+/+$ ♀ $\times$ $R/+/+$ ♂ gives a 1:2 ratio of colored and colorless kernels (1282 to 2451 in a series of actual crosses). Again, this is the ratio expected with random segregation and elimination of all n $+$ 1 pollen grains.

Correlating Linkage Groups and Chromosomes.—Trisomic types provide a useful general method of correlating specific genes with specific chromosomes. The example chosen for the above discussion was originally used as a demonstration that the R gene and other genes in the same linkage group are carried in the smallest of the ten chromosomes of maize, the tenth chromosome. A number of other linkage groups in maize were first correlated with specific morphological chromosomes by this means. The method is also generally useful in localizing new mutations, particularly in plants where crossing over occurs in both sexes. It has an obvious advantage over the ordinary linkage method in that it is independent of the amount of crossing over. As an illustration of this, let us assume a hypothetical gene x in maize. In locating this gene by means of the ordinary linkage method, test crosses would be made that involved gene x and genes in known linkage groups. If gene x happened to lie in chromosome 10, it would show readily detectable linkage with the R gene only if it happened to lie sufficiently close to R to give appreciably less than 50 per cent of recombinations with it. In other words, the fact that gene *x* is in chromosome 10 might easily be missed unless it were tested with two or three known genes favorably located in chromosome 10. But by making the cross of a triplo-10 plant with a diploid carrying *x*, getting a trisomic plant of the constitution $+/+/x$, and using pollen of this on the diploid *x/x*, one can tell by the ratio obtained whether or not *x* is in chromosome 10. A 1:1 ratio for $+$ and *x* tells one immediately that *x* is not in chromosome 10, a ratio of 2:1 for $+$ and *x*, that it is in this chromosome. Its location within the chromosome can then be determined by the standard testcross method.

Tetrasomic Types.—*Tetrasomic* types, individuals with a given chromosome represented four times, are possible for some chromosomes in some plants and animals. They are sometimes obtained in plants by self-pollinating a trisomic individual or by crossing two such plants. Their initial zygotic frequency will be determined, of course, by the frequency with which the particular extra chromosome is transmitted through the gametes. Tetrasomic individuals have the phenotypic characteristics of the corresponding trisomics in exaggerated form. Consequently they are regularly weaker and less viable than the trisomic types, involving the same chromosome. From the standpoint of genic balance it is obvious that they deviate from the normal in the same direction but to a greater degree than do comparable trisomic types.

Monosomic Types.—It is evident that the type of segregation that gives rise in the first instance to a trisomic type will also give rise to a spore or gamete with a corresponding chromosome absent. From the fact that regional deficiencies are usually eliminated in the haploid generation in plants, a gametophyte deficient in an entire chromosome would not be expected to survive; in fact they do not in diploid plants. In certain species of plants, individuals with one less than the normal number of chromosomes are known to occur, but in such species it is also known that more than two sets of chromosomes are present. We shall therefore postpone discussion of so-called monosomic types in plants until the general question of polyploidy has been taken up.

THE FOURTH CHROMOSOME OF DROSOPHILA

Both trisomic and *monosomic* (2n minus 1—usually written 2n — 1) types are known in Drosophila. These are readily obtainable for the small fourth chromosome but are inviable when they involve the large second or third chromosomes. Because of their relation to sex, the X chromosomes present a somewhat special case which we shall discuss later in this chapter. Flies trisomic for chromosome 4, usually referred to as triplo-4 individuals, are only slightly different from normal diploids in phenotype—one cannot depend on identifying them by their appearance. The relation is well illustrated by the behavior of the fourth chromosome character eyeless, differentiated by the recessive gene *ey*. The cross $+/+/ey$ ♀ $\times$ ey/ey ♂, or its reciprocal, gives approximately the 5 $+$ to 1 *ey* ratio expected with random segregation and no chromosome elimination or gamete competition (Fig. 91). In this case there is no crossing over between the eyeless locus and the centromere. With no mortality the cross $+/+/ey \times +/+/ey$ should give a phenotypic ratio of 35 $+$ to 1 *ey*. The tetra-4 individuals die, however, and the observed ratio approximates the 26:1 ratio expected on this basis.

Non-random Segregation.—The above discussion assumes that the segregation of the three homologs in a triplo-4 fly is entirely random. While this is sometimes true it is not necessarily so. Sturtevant has found that there are systematic deviations from randomness in the segregation of fourth chromosomes in triplo-4 females and, furthermore, that individual fourth chromosomes (*i.e.*, those derived from a common source) have specific and predictable properties with respect to their behavior in combinations with other fourth chromosomes. The significance of the rather surprising relation found here is not fully understood.

Haplo-4 in Drosophila.—Monosomic-4 or haplo-4 individuals of Drosophila differ phenotypically from wild-type in a characteristic way. They are smaller, weaker, develop more slowly, have slightly roughened eyes, and have short slender bristles (Fig.

92). A comparison of the triplo-4 and haplo-4 types provides a good illustration of the genic balance conception. The numerical proportion between fourth chromosome and other chromosomes in a triplo-4 fly is 3:2 or, expressed as an index, 1.5, while that in a haplo-4 is 1:2 or 0.5. In other words, in terms of these indices, wild-type is 100 per cent greater than haplo-4, but triplo-4 is only 50 per cent greater than wild-type. It should be pointed out, however, that these simple relations almost certainly do not tell the whole story. They do, nevertheless, represent a general rule; the phenotypic effects of a specific duplication are less marked than are those of the complementary deficiency.

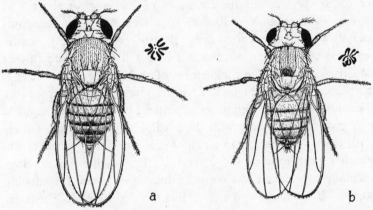

Fig. 92.—Drosophila melanogaster. a, wild-type; b, haplo-4. Oogonial metaphase chromosome groups of such individuals are also shown. (From Morgan, "The Theory of the Gene," Yale University Press.)

The deficiency of an entire fourth chromosome in haplo-4 is roughly equivalent to a deficiency in a segment of equal length in one of the other three chromosomes. The genetic effects are correspondingly similar. A cross of a diploid fly carrying any fourth chromosome recessive, to a haplo-4 individual, results in the appearance of the recessive character in the F_1 generation in all haplo-4 zygotes. In such instances it often happens that the recessive character is more extreme than in a homozygous recessive diploid. This illustrates another general rule applicable to sectional deficiencies as well as to whole chromosome deficiencies—

in combination with any allele of a specific gene, a deficiency gives a phenotypic effect at least as extreme as that of the most extreme known allele of the gene concerned.

NON-DISJUNCTION OF THE X CHROMOSOMES

Turning to the rather special case of the X chromosome of Drosophila, it is found that about one egg in 2,500 receives two instead of one X chromosome from the mother. A non-disjunctional egg of this type may be fertilized by either an X or a Y sperm. The former gives a triplo-X individual, the latter an XXY female. The frequency of XXY exceptional females will, of course, be one-half the frequency of XX eggs or about 1 in 5,000. The triplo-X individual is a weak sterile female which usually dies before emergence. Because of its bearing on the question of sex determination this type will be discussed further in Chapter XVI. The complement of an XX egg is a no-X egg. Such an egg, fertilized by a Y sperm, gives rise to a zygote that dies very early in the egg stage. Fertilized by an X sperm, a no-X egg gives an XO male with the sex-linked genes of the father. This type is comparable to a monosomic type, but, because of the peculiar properties of the X and Y chromosomes, is developmentally very similar to a normal male. It is, however, invariably sterile. It should be emphasized that the sterility of XO males in Drosophila is not a general rule for all animals. It will be recalled that the XO condition is usual in the males of some species. The frequency of no-X eggs is about one in 600 (this will mean a frequency of exceptional males of about one in 1,200), or about four times that of XX eggs. This difference is evidently due to the fact that failure of one of the X chromosomes to be included in a daughter nucleus at the first division is a more frequent deviation from normal than is the inclusion of both X's in a single daughter nucleus.

Detection of Non-disjunctional Types.—These exceptional matroclinous females and patroclinous males, known as non-disjunctional exceptions, are of course readily detected genetically if the female that gives rise to them is homozygous for sex-linked recessives for which the male parent carries a dominant allele. Thus in the mating of w/w ♀ $\times$ $+$ ♂, the normal daughters are $+/w$ and the normal sons are w. The non-disjunctional daughters will be w/w, or white eyed like the mother, and the exceptional sons will be $+$, or red eyed like the father. The exceptional daughters, unlike normal females, will have a Y chromosome.

$$XXY\,♀ \qquad X \qquad XY\,♂$$

	SPERMS WITH RELATIVE FREQUENCIES	
	X 1	Y 1
EGGS WITH RELATIVE FREQUENCIES		
XX 2	XXX 2 USUALLY DIES	XXY 2 EXCEPTIONAL ♀
Y 2	XY 2 EXCEPTIONAL ♂	YY 2 DIES
XY 23	XXY 23 REGULAR ♀	XYY 23 REGULAR ♂
X 23	XX 23 REGULAR ♀	XY 23 REGULAR ♂

Fig. 93.—Diagram showing the offspring produced by an XXY female (*i.e.,* secondary non-disjunction).

That they are females without detectable phenotypic differences from normal in spite of the presence of the Y, demonstrates again the relatively passive rôle played by the Y chromosome.

Secondary Non-disjunction.—If an exceptional white eyed female obtained as described above is again mated to a normal male it is found that about 4 per cent of her daughters are exceptional, *i.e.*, *w/w*, and that likewise about 4 per cent of her sons are exceptional. The frequency of non-disjunction is many times higher than that in normal females. This increase is due to the presence of the Y chromosome. In contradistinction to non-disjunction in an XX female, which is called *primary non-disjunction,* this type of segregation in an XXY female is known as *secondary non-disjunction.* The types obtained from a white eyed XXY female mated to a normal wild-type male are indicated in figure 93. In addition to the types already discussed it is seen that one new type appears, namely, an XYY male. This has been identified cytologically, and is known to be a normal male in no way distinct from an XY male, except for the presence of the extra Y chromosome which is transmitted to the descendents as expected. The interpretation has been checked further by showing by breeding tests that all exceptional daughters are XXY and that, among regular daughters, XX and XXY individuals occur with approximately equal frequencies. The presence of XXY, XXX, as well as the XYY male already mentioned, have been demonstrated by Bridges by direct cytological examination.

Frequency of Secondary Non-disjunction.—Breeding tests of secondary exceptional females from XXY mothers which were heterozygous for several sex-linked genes show that the X chromosomes that go together to one pole are practically always non-crossovers. Regular offspring, on the other hand, show approximately the normal frequency and distribution of crossovers. These facts tell us at once that crossing over must be intimately related to disjunction in XXY females. In order to discuss further the nature of secondary non-disjunction, it will be convenient to make use of simple formulae for relating gametic and zygotic frequencies of non-disjunctional types. This is necessary because of the fact that XX and Y eggs are not recovered as viable zygotes when fertilized by X and Y sperm respectively. Letting p equal the proportion of XX-Y segregation (non-disjunction for the X's), and q the proportion of exceptional types among viable zygotes, it can be seen by examination of the relations shown in figure 93 that $q = \dfrac{p}{2 - p}$ and that $p = \dfrac{2q}{1 + q}$.

Examining the consequences of random formation of a bivalent (XX and Y or XY and X) at metaphase, we see that the frequency of these will be ⅔ XY and X to ⅓ XX and Y. The XY conjugation will give XX-Y and XY-X segregation with equal frequencies while XX conjugation will give XY-X segregation only. The proportion of XX-Y segregation (p) on this basis will then be 0.33 and the frequency of exceptional zygotes will be 0.2 $\left(q = \dfrac{0.33}{1.67} = 0.2\right)$ or 20 per cent. From this we might assume that XY conjugation occurs enough less than two-thirds of the time to account for the observed frequency of non-disjunction. As will become evident from considerations discussed below, however, this hypothesis is inadequate to explain the facts.

Inversions and Secondary Non-disjunction.—An XXY female in which one X chromosome carries an inversion shows an increase in the frequency of secondary non-disjunction over controls homozygous for the normal sequence. The magnitude of this increase varies with different inversions. The inversion known as "delta-49" (an arbitrary designation) gives a frequency of non-disjunctional zygotes of about 45 per cent, a value higher than for any other inversion so far studied. This corresponds to a gametic frequency of about 62 per cent $\left(p = \dfrac{0.90}{1.45}\right)$. Obviously this is higher than is possible with random formation of a bivalent and a univalent. With XY conjugation occurring with a frequency of 100 per cent, followed by random segregation of the univalent X, XX-Y segregation would be expected 50 per cent of the time. This would mean a frequency of zygote exceptions of 33.3 per cent—less than the observed value for an XXY delta-49 heterozygote. A calculation of the frequency with which non-exchange X chromosome tetrads occur in such an XXY delta-49 heterozygote indicates that most, if not all, of these must give XX-Y segregation to account for the 45 per cent of observed exceptional zygotes. The point of all this is that segregation is not random but must somehow be closely related to crossing over. The assumption that exchange X-tetrads in XXY females always result in XY-X segregation, and that non-exchange XX tetrads in such individuals always result in XX-Y segregation is consistent with the known facts. Why this should be so, if it is—and it should be emphasized that the assumption may be incorrect— remains unexplained. The general problem of the mechanism of secondary non-disjunction is unsolved.

Aberrations as Proof of Chromosome Theory.—The several examples of aberrant chromosome behavior discussed in this chapter are of historical importance in that they provided convincing proof of the correctness of the chromosome theory of heredity. Non-disjunction of the X chromosomes of Drosophila, for example, was deduced from purely genetic results at a time when many biologists were loath to accept the chromosome theory. Bridges' direct cytological observations of the predicted chromosome constitutions served to remove most of the remaining doubts as to the correctness of this theory.

REFERENCES

Blakeslee, A. F. 1934. New Jimson weeds from old chromosomes. Journ. Hered., 25:80–108.

Bridges, C. B. 1916. Non-disjunction as proof of the chromosome theory of heredity. Genetics, 1:1–52, 107–163.

Catcheside, D. G. 1937. The extra chromosome of Oenothera lamarckiana lata. Genetics, 22:564–576.

McClintock, B., and H. E. Hill. 1931. The cytological identification of the chromosome associated with the R-G linkage group in Zea mays. Genetics, 16:175–190.

PROBLEMS

1. The twelve trisomic types in Datura stramonium are given names. The "pointsettia" type has three homologs of the chromosome that carries a gene differentiating purple and white flowers (w—white is recessive). Assuming that the extra chromosome is transmitted to 50 per cent of the offspring through the female gametophyte and to none through the pollen, and that there is no crossing over between w and the centromere, what ratios would the following give:

    (a) $+/+/w$ $\times$ w/w

    (b) w/w $\times$ $+/+/w$

    (c) $+/w/w$ $\times$ w/w

    (d) w/w $\times$ $+/w/w$

(Note: The symbol w used here is not the usual one used for this gene. It is used here for the sake of consistency.)

2. Making the same assumptions regarding transmission of the extra chromosome as for Datura, what ratios would you expect in maize from the following crosses involving trisomic 9:

    (a) $+/+/wx$ $\times$ $+/+/wx$

    (b) $+/+/wx$ $\times$ $+/wx/wx$

    (c) $+/wx/wx$ $\times$ $+/wx/wx$

What ratio would you expect in the mature pollen from a $+/+/wx$? (All pollen, n and $n + 1$, appears to be normal.)

3. With a transmission ratio of 30 per cent for trisomic 9 of maize through the female gametophytes (*i.e.*, 30 per cent of the off-spring of the cross trisomic $\times$ normal are trisomic), what ratio of $+$ to wx would you expect from:

    (a) $+/+/wx$ $\times$ wx/wx

    (b) $+/+/wx$ $\times$ $+/+/wx$

    (c) $+/wx/wx$ $\times$ wx/wx

(Note: Crossing over between wx and the centromere is so infrequent that it may be disregarded.)

4. Having available 12 strains of Datura in each of which there is a different one of the possible trisomic types present, outline how you would go about determining the chromosome in which the locus of a new recessive mutation is carried.

5. By means of a checkerboard, work out how a 26:1 ratio is obtained in Drosophila by crossing two triplo-4 individuals of the constitution $+/+/ey$.

6. What ratio of triplo-4 to diplo-4 to haplo-4 would you expect in Drosophila from the mating triplo-4 female by haplo-4 male?

7. XXY females of Drosophila of a certain constitution gave 40 per cent of non-disjunctional zygotes. What percentage of non-disjunctional gametes does this correspond to?

8. Starting with XXY females of Drosophila in which each X chromosome carries a white allele, how would you go about getting XXY females homozygous for the eosin (w^e) allele of the white

gene? From these how would you proceed to get XXY females homozygous for the two sex-linked genes scute (*sc*) and miniature (*m*)?

9. By means of genetic tests, how can XY and XYY males be distinguished from each other?

10. How would you correlate the fact that the X chromosome of D. pseudoobscura is about 180 map units long with the fact that XXY females in this species give almost no non-disjunctional gametes?

CHAPTER XVI

THE DETERMINATION AND DIFFERENTIATION OF SEX

Heterozygous Male Type.—The type of sex-determination that depends on an unpaired chromosome, or an unequal pair, in the male has been discussed in earlier chapters as occurring in man and in Drosophila. This same type is characteristic of mammals in general, of many orders of insects (*e.g.,* Orthoptera, most Hemiptera, Coleoptera, Diptera), of nematodes, of some fish, and of several other groups of animals, as well as of a few seed-plants.

Fig. 94.—Feather from a Barred Plymouth Rock fowl.

Heterozygous Female Type.—In some groups—birds, moths and butterflies, some fish, and perhaps other forms—essentially the same system is found, except that here the unequal pair occurs in the female. One of the best known cases demonstrating this relation is that of the inheritance of the barred pattern in fowls. The familiar Barred Plymouth Rock breed has feathers with successive transverse bars of black and white (Fig. 94). This pattern is due to a dominant gene, *B*. If crosses are made between a barred and a black strain, the following results are obtained:

P — barred ♀ × black ♂   black ♀ × barred ♂

F₁ — black ♀ ♀, barred ♂ ♂   barred ♀ ♀, barred ♂ ♂

F₂—{ 1 barred ♀   1 barred ♂   { 1 barred ♀   2 barred ♂
    { 1 black ♀   1 black ♂   { 1 black ♀

It is clear that these results are like the ones to be expected from a dominant sex-linked gene in Drosophila—except that the sexes are reversed in every case. The interpretation is in agreement with this. The female is found cytologically to have an unequal pair of chromosomes (Fig. 95), and the larger member

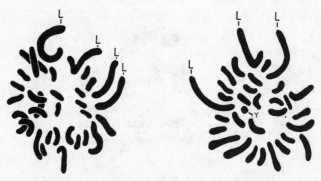

Fig. 95.—Diploid chromosome groups of the fowl. Male at left, female at right. The Y is the smallest chromosome; the long chromosomes labelled "L" include a pair of autosomes and two X's (in the male) or one (in the female). (After Shiwago.)

of this pair is represented twice in the male. This large chromosome is sometimes called X, the smaller mate, Y; another convention is to use the letters Z and W. The choice between these is a matter of taste; the former seems to us rather less confusing.

The formulae then are: ♀ = XY, ♂ = XX. In the cross shown above, each F₁ female gets her single X from her father, and this determines whether she is barred or black—*i.e.*, the gene *B* lies in the X chromosome. Since the F₁ males all get an X from each parent, their barred plumage is due to the dominance of *B* over the + allele present in the black strain. Analysis shows that the F$_2$ results are in agreement with this interpretation. It will be noted that one can conclude that the Y is without influence here; this is a general rule in sex-linked inheritance in birds and Lepidoptera, just as is the similar conclusion with respect to the Y in insects and mammals—though exceptions are known in both types.

Historical Relations.—There is a curious historical relation here. The Drosophila type of sex-determination was discovered first, and was demonstrated cytologically in a wide variety of animals before the corresponding type of sex-linkage was recognized; the bird type of sex-linkage was found before the corresponding chromosome picture, as the table shows:

	Cytological demonstration	Sex-linkage
Drosophila type	Stevens, 1905	Morgan, 1910
Bird type	Seiler, 1913	Doncaster and Raynor, 1906

Thus, from 1906 to 1910, one type was known cytologically, and was known to occur in many different groups of animals, while the other type was known through sex-linkage in birds and moths and was therefore also supposed to be of general occurrence. This apparent contradiction came at a time when the chromosome interpretation of heredity was just beginning to develop rapidly, and there can be no doubt that this development was greatly retarded by the skepticism aroused by such a glaring discrepancy.

Heterozygous Sporophyte Type.—There is another type of sex-determination that rests on the segregation of an unequal pair of chromosomes. In this type, found in the Bryophytes (mosses, liverworts), the conspicuous phase, in which the sex is developed, is the haploid one—the gametophyte. Male and female plants differ in a single chromosome (Fig. 96); both are present in the fertilized egg. Meiosis leads to the formation of four haploid spores; of these, two have the large chromosome ("X") and develop into females, while two have the small one ("Y") and develop into males. Here, as in some other cases already referred to, it is possible to demonstrate directly that segregation occurs at the meiotic divisions.

These three systems of sex-determination—heterozygous male, heterozygous female, and heterozygous sporophyte—are the only

Fig. 96.—Chromosomes of Sphaerocarpos. Female at left; the largest chromosome is the X. Male at right; the smallest chromosome is the Y. (After Allen.)

ones that regularly and automatically lead to the production of equal numbers of males and females. There are, however, several other mechanisms, which give variable sex-ratios.

HAPLOID MALE TYPE

One such system is that in which the female is diploid, the male haploid. This occurs in the Hymenoptera (bees, wasps, ants, sawflies, etc.), thrips, a few of the Homoptera (white-flies, some scale-insects), at least some of the mites, and probably in rotifers. The first example studied in detail was that of the honey-bee, for which "Dzierzon's rule" was propounded many years ago: fertilized eggs give rise to females, unfertilized ones to males. As applied to Hymenoptera in general, and apparently even to some races of honey-bees, this rule requires some revision; for females may at times arise from unfertilized eggs—there are species in which males are almost completely absent, and others in which they occur only in alternate generations. Parthenogenetic development may occur after the formation of only one polar body and no reduction in chromosome number (giving rise to a diploid female), or with two polar bodies and chromosome reduction (giving rise to a haploid male). The former occurs only rarely in the honey-bee, so that Dzierzon's rule is, for all practical purposes, valid there.

Meiosis in Haploid Males.—The haploid males produced in the forms under discussion give rise to haploid sperm. Meiosis is unable to proceed as in a diploid, since the chromosomes are not present in pairs. There are usually two divisions like the meiotic ones; but at the first one the nucleus does not divide, and a nonnucleated bud is pinched off from the cell (Fig. 97). The second division of the nucleated cell is usually a normal equational division, resulting in the formation of two haploid spermatids which differentiate and thus give rise to two functional sperms from each primary spermatocyte. (The honey-bee is exceptional

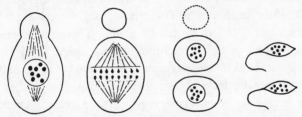

Fig. 97.—Diagram of meiosis in a haploid male Hymenopteron. The first division occurs with the nuclear wall still intact, and results in the pinching off of a fragment without a nucleus. The second division is a normal equational one. The time at which chromosome division occurs is not definitely known; the figure arbitrarily shows it as happening between the metaphases of the first and second divisions.

in that the second division results in one functional and one abortive—though nucleated—spermatid).

In most Hymenoptera, eggs usually go through two polar body divisions, and thus contain haploid nuclei. It is certain that such an egg, if unfertilized, may develop into a male; if it is fertilized the diploid number is restored and a female develops.

Genetic Results of Haploid Male Condition.—It will be seen that in these animals all the chromosomes have the same kind of descent as do the X's of Drosophila or of man; every male gets *all* his chromosomes from his mother (since he has no father), every female produced from a fertilized egg gets a set of chromosomes from each parent. All genes accordingly show the same type of transmission as do the sex-linked genes of Drosophila. This has been shown especially by Whiting and his co-workers, who have studied the wasp-like parasite, Habrobracon.

Genetic Basis of Haploid Male Type.—The difference between male and female in Drosophila depends on the fact that one has a single X, the other has two. Is the sex of Hymenoptera, then, determined in a similar way? We shall see later in this chapter that there is reason for suspecting that a haploid Drosophila would be female; and Whiting has shown that it is possible to obtain males of Habrobracon that are diploid throughout; we are, therefore, forced to conclude that the type of sex-determination is essentially different in the two groups.

The interpretation to be given to the occasional diploid males found in Habrobracon is still uncertain; but the most probable view seems to be that suggested by Snell. There are a series of independent loci, in different chromosomes, say A, B, C, D, etc., of such a nature that heterozygosis in any one or more of them produces a female; lack of such heterozygosis (*i.e.,* either haploidy or complete homozygosis) produces a male. There are enough of these loci, and ordinary strains carry separate alleles in enough of them, so that the chance of getting a complete homozygote (*i.e.,* a diploid male) is extremely small. Inbreeding automatically decreases the heterogeneity; and the fact is that diploid males occur in appreciable numbers only after inbreeding has been practiced.

Diploid males have not yet been found in the other organisms in which the males are normally haploid; it remains uncertain whether the Habrobracon system applies to them or not.

HERMAPHRODITISM

Bonellia.—Another type of sex-determination is illustrated by the marine worm-like animal Bonellia (Fig. 98). In this form the male is about $\frac{1}{500}$ the length of the female, and during the sexual phase of its life lives in the uterus of the female. Here there appears to be no genetic control of sex; the embryos are free-swimming organisms, and their sex is not determined. The decisive event occurs when a free-swimming larva settles down and undergoes further development. As shown by Baltzer, a larva that happens to settle down on the proboscis of a female develops

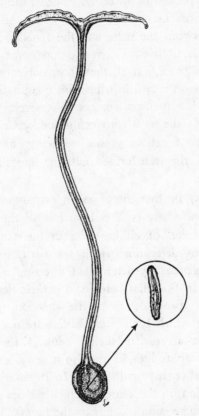

Fig. 98.—Bonellia. Female at left, with male in the uterus. At right, male, greatly magnified. (After Baltzer and others.)

into a male and later migrates to the uterus of the female; a larva that settles down by itself, on the sea bottom, develops into a female. That is, sex is determined by extrinsic factors, not by genetic differences among the larvae. These larvae may be thought of as potential hermaphrodites; in any case, each one has the necessary genetic constitution that enables it to develop into either sex, given the appropriate external stimulus.

Sex Reversal.—In the usual Drosophila or bird type of sex-determination, it is evident that each sex has present in it all the genes that are present in the other. Since it can be shown that the Y is not essential (at least in many species), the only significant differences between the male and the female chromosome complexes are quantitative—a given chromosome (X) is present either once or twice. It is, therefore, not surprising that occasional cases are on record in which this quantitative difference has been overruled. In fowls, for example, several individuals have been described (the most convincing one by Crew) in which an individual, after functioning and producing offspring as one sex, has then gradually transformed into the opposite sex and functioned as such.

Bonellia may be thought of as an extreme case of the more usual condition, where each embryo has all the genes necessary for the production of either sex. In the familiar forms the dosage relations determine which sex shall develop, and only rarely is this genetically determined direction overridden by external agents; in Bonellia there is no genetic determination, and external agents always determine the direction. There are many other forms in which both potentialities are not only present in the embryo, but are realized in the adult. Most seed-plants are functional hermaphrodites, as are also many animals—*e.g.*, earthworms, sponges, many snails, etc. In these forms there is no sex-determination, in the sense in which we have been using the term, because there is only one sex—the hermaphrodite.

Sex in Maize.—Maize is a hermaphroditic (*i.e.*, monoecious) species, the anthers in the tassel being male organs and the pistils in the ear female ones. There exist mutant genes that alter these relations: *ba* (barren stalk) is a recessive that eliminates the ears, thus making *ba ba* plants males; Ts_3 and ts_2 ("tassel-seed"—one dominant to the normal condition, the other recessive) cause the tassels to produce pistils instead of anthers. The following types may be produced:

$+\ ts_2\ ba\ ba$ — normal tassels, no ears—male
$ts_2\ ts_2\ ba\ ba$ — pistillate tassels, no ears—female

$+\ +\ ba\ ba$ — normal tassels, no ears—male
$Ts_3\ +\ ba\ ba$ — pistillate tassels, no ears—female

Either of these pairs produces fertile strains, that breed true to the dioecious condition. That is, a hermaphroditic plant has been transformed to a dioecious one by the introduction of two mutant genes; also, there are two ways of doing this, of which one gives heterozygous males, the other heterozygous females.

Fish.—It is possible that a somewhat similar situation exists in certain fishes, where it has been shown that closely related forms may differ in respect to whether the male or the female is heterozygous for sex-determining genes. It should, however, be pointed out that the maize example fails in one respect to furnish a complete parallel to the situation found in Drosophila and in fowls. In the latter forms the heterozygous sex is effectively haploid for the chromosome pair concerned, the Y chromosomes being of no importance; in the maize examples, the analogues of the Y are dominant genes on which the whole stability of the system depends. There have been many speculations concerning the origin of the Drosophila and the bird systems, and their historical relationship to each other. The origin of the "inert" Y has likewise been much discussed in connection with these questions. No clear solution of the questions involved is yet apparent; but these artificial dioecious strains of maize at least serve to suggest some possibilities.

INTERSEXES

In many dioecious species strains have been found that produce specimens, called *intersexes,* that are intermediate between typical males and females. These are not functionally hermaphroditic, and they differ from gynandromorphs in that they are not made up of a mixture of typically male and female parts; they are true intermediates. Study of the inheritance of intersexuality and of the development of intersexes has contributed much to current knowledge of the mechanism of sex-differentiation.

Intersexes in Drosophila.—Perhaps the clearest case is that which Bridges has studied in Drosophila melanogaster. Here the intersexes appear among the offspring of *triploid* females— *i.e.,* females that possess three complete sets of chromosomes. Such females are very similar to diploid females, but with practice may be distinguished by their more robust appearance, heavier bristles, and (as Dobzhansky has shown) by the larger size of the cells resulting in more sparse distribution of the minute hairs on the surface of the wing. That such females are in fact triploid (usually denoted as 3n, n being the designation for one complete haploid set of chromosomes) has been shown by cytological study. Their origin is to be referred to the occurrence, in a diploid female, of a chromosome division without a separation of the products into two nuclei. Such a process, occurring in the germ-line at a mitotic division, will give rise to an island of tetraploid (4n) tissue, from which will arise diploid mature eggs as a result of meiosis; if the aberrant division is a meiotic one a diploid egg will result directly. If such a diploid egg is fertilized by an X-bearing sperm, the result will be a triploid female. Such a series of events does not occur with a predictable frequency; but new occurrences of triploids from diploid parents have been detected many times.

Segregation in Triploids.—Triploid females lay many eggs; when they are mated to normal diploid males a great many of these eggs contain embryos that die in early stages, but the remaining ones develop into a variety of adult offspring. Study of these viable types shows, as will appear below, that meiosis in the triploid results in equal numbers of eggs with two X's and with one, with two second chromosomes and with one, etc.—just as in the case of the trisomic types already described. The separate chromosomes do not all behave in the same way in any one cell— *i.e.,* an egg may receive two X's and only one second chromosome. No offspring are produced with one second and two third chromosomes or with two seconds and one third; there can be no doubt that such eggs are formed, and give rise to a large proportion of the inviable zygotes that are known to be present. Neglecting the dot-like fourth chromosome for the moment, the mature eggs that give viable zygotes are of four kinds: designating one second and one third chromosome by the symbol A (for autosomes), these four types may be written:

$$2X \ 2A$$
$$1X \ 1A$$
$$2X \ 1A$$
$$1X \ 2A$$

The two types of segregation concerned here do not occur equally frequently, that producing the third and fourth types of eggs being about twice as frequent as that producing the first and second. The reason for this inequality is not known; it is convenient to remember that the more frequent type is the one that gives more nearly equal numbers of chromosomes at the two poles.

Offspring of Triploids.—The various zygotic types produced when these eggs are fertilized by the sperm of a normal diploid male are shown in the following table:

	Sperms	
Eggs	1X 1A	1Y 1A
2X 2A	3X 3A, triploid female	2X 1Y 3A, intersex
1X 1A	2X 2A, diploid female	1X 1Y 2A, male
2X 1A	3X 2A, super-female	2X 1Y 2A, diploid female
1X 2A	2X 3A, intersex	1X 1Y 3A, super-male

All these types have been identified both genetically and cytologically.

The triploid females are like their mothers, both phenotypically and in genetic behavior, and may be used to continue the strain. By the use of mutant genes it is possible to demonstrate genetically that each of them receives two whole sets of chromosomes from her mother, one from her father. The two kinds of diploid females, one with no Y and with one X from each parent, the other with a Y and with two X's both derived from the triploid mother, may be distinguished genetically by the use of mutant genes; *e.g.*, if the father is bar, the mother not-bar, the first type will have broad (*i.e.*, heterozygous) bar eyes, the second will be wild-type.   Genetic tests show, as expected, that the second type gives secondary non-disjunctional offspring (due to the presence of a Y); the first type does not.   The diploid males are normal,

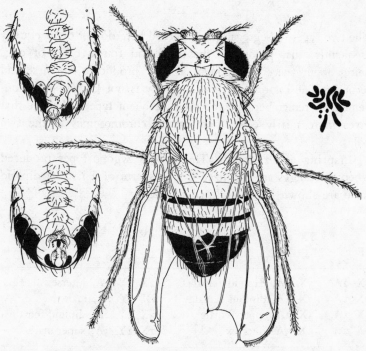

Fig. 99.—Triploid intersex of Drosophila melanogaster.   At left, ventral surfaces of abdomens of female-like (above) and of male-like (below) inter-sexes.   At right, metaphase chromosome group of an intersex.   (After Bridges, from Morgan, "The Theory of the Gene," Yale University Press.)

both phenotypically and genetically.   It can be shown that each receives his X from his mother.

Intersexes.—The remaining four types of individuals are the ones of most interest in the present connection.   All of them are sterile, but their constitutions have been determined both by cytological study and by the use of mutant genes as markers.   The two forms with the formula 2X 3A, one with a Y, the other without, are phenotypically indistinguishable. They are not all alike, but all are typical intersexes—having both male and female characteristics (Fig. 99). The "sex-combs" on the front leg are usually present (normally present in males, absent in females); the gonads are usually either rudimentary testes or ovaries, sometimes intermediate organs, sometimes one ovary and one testis; the ducts and external genitalia may be either male or female, or may be mixtures of the two. Examination of the chromosome formula for these specimens shows that they have the two X's characteristic of normal diploid females, but differ from them in having an extra set of autosomes. The autosomes are the same as those of triploid females, but there is one less X present. These relations indicate that the relative amounts of X and autosomes are responsible for the intersexual phenotype. One may surmise that the X influences development in the female direction, the autosomes in the male direction—which is evidently consistent with the fact that addition of a single X to the composition of a normal diploid male gives the formula for a diploid female.

Ratio of X to Autosomes.—Assuming that the effective element is the ratio betwen the number of X's and the number of sets of autosomes, we have the following relations:

Type	Formula	Ratio X/A
Super-female	3X 2A	1.5
Triploid female	3X 3A	1.0
Diploid female	2X 2A	1.0
Intersex	2X 3A	0.67
Male	1X 2A	0.5
Super-male	1X 3A	0.33

Supersexes.—The table shows the reasons for the names given to the types super-females and super-males. Both of these are sterile and poorly viable types (we have already referred to the super-females in connection with non-disjunction—they are most easily obtained from attached-X mothers). They have the essential structures, both internal and external, of females and males, respectively (Fig. 100). They are, in fact, super-sexes by composition rather than by phenotype. Their structures do not, however, contradict the general scheme just outlined.

Other Types.—A few other types are also known, being obtainable by special and rather laborious methods. Occasionally tetraploid females ($4n = 4X\ 4A$) are produced; they are fertile and phenotypically nearly normal females—as expected from their X/A ratio of 1.0. The two types $4X\ 3A$ and $3X\ 4A$ have been seen; the first resembles superfemales (X/A = 1.33), the second is an intersex (X/A = 0.75), but too little is known about them to judge as to whether they are phenotypically distinguishable from other super-females and intersexes, respectively,

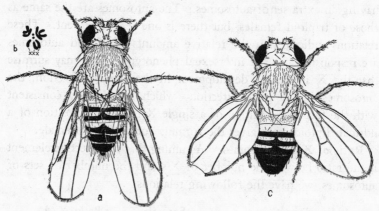

Fig. 100.—a, superfemale of Drosophila melanogaster, with chromosome group (b); c, supermale. (From Bridges.)

on the basis of their sexual characters alone. All these types are consistent with the rule: If the ratio X/A is 1.0 or greater, the individual is a female; if it lies between about 0.6 and 0.8, the individual is an intersex; if it is 0.5 or less the individual is a male.

It is not likely that the effective relation between X's and autosomes is as simple as implied in the use of the ratio, but as a first approximation this relation does serve to interpret the results. It will be seen that it leads to the expectation that a haploid individual would be female ($1n = 1X\ 1A$, X/A = 1.0). There is some evidence suggesting that haploid sectors of mosaics are in fact female; but this evidence is not wholly convincing. The question can not be considered finally settled.

Number of Sex Genes.—The analysis has been carried through in terms of X's and sets of autosomes as whole units. It remains to examine the question, whether two or many genes are concerned. One limitation has already been made; we have specified that "A" is to be taken as meaning one second and one third chromosome, the dot-like fourth being neglected. This is because most triploid strains have only two fourth chromosomes, the third one that was originally present being usually lost by chance after a few generations. The indications are that it makes little difference to the phenotype or fertility of a triploid female, whether she has two or three fourth chromosomes; apparently, though less certainly, the phenotype of an intersex is likewise unaffected.

In the case of the X, the results of Dobzhansky and Schultz show that many genes are concerned. They studied the effects of adding, separately, a series of duplicating fragments of X chromosomes to the 2X 3A constitution of intersexes. Any such fragment, from any part of the X, appears to shift the grade of intersexuality toward femaleness; and the extent of the shift is roughly proportional to the size of the fragment. The X, therefore, carries a series of different genes that influence development in the direction of femaleness; presumably it also carries genes working in the male direction, but the balance of any section is likely to be toward femaleness. Less direct evidence suggests that there are also many "male" genes in the autosomes. These results, then, lead to the conception that the individual is the resultant of a balance among the effects of a whole series of genes, which, individually, influence development in many different directions.

Turning Point.—Dobzhansky and Bridges have studied the development of the intersexes of Drosophila. They find that intersexes start development as males; at a given moment (perhaps not the same moment in every part of an individual) the development shifts to that characteristic of the female. Characters determined before this *turning point* are male, those determined after it are female. The grade of intersexuality then depends on the time at which the turning point occurs; if it is early the intersex is female-like, if it is late the intersex is male-like.

Intersexes in Plants.—Triploid intersexes similar to those in Drosophila are known also in the seed-plant Rumex, and in several Lepidoptera. In the latter case, since the female is the XY sex, triploids with three X's are males; otherwise the system is similar to that in Drosophila, so far as it is known. In mosses and liverworts diploid intersexes occur. The normal sexes are haploid—♀ = 1X 1A, ♂ = 1Y 1A, intersex = 1X 1Y 2A. It is not known what rôle the autosomes play here.

Intersexuality in Lymantria.—Diploid intersexes are known in a number of animals. In some of these they are due to simple gene mutations. The best known example, however, is that found in the gipsy-moth, Lymantria dispar. This case, studied by Goldschmidt, is the one on which were originally based the term intersex, the idea of opposing male and female tendencies in development, and the demonstration of a turning point. In Lymantria the intersexes are produced as a result of crossing certain geographical races. There are two opposing genes or sets of genes, F (female producing) and M (male producing). These differ in "strength" in different races, and certain combinations result in intersexes. The genetic interpretation to be put on the concept of "strength" is uncertain—probably a series of multiple alleles is concerned, but modifying genes play a part that cannot yet be accurately estimated. "M" is carried in the X chromosomes; the inheritance of "F" is uncertain. Goldschmidt shows that his experiments, designed to test this point, give contradictory results. In view of these uncertainties the case is not satisfactory for fuller description here.

Sex Differences and Development.—In some respects the study of the determination and development of sex appears to be the most hopeful way of analyzing the effects of genes on development; for the genetic basis is well understood, and the anatomical, embryological, and physiological knowledge of sex is perhaps more highly developed than is such knowledge for any other character. Unfortunately, however, the genetic situation is complex; male and female do not differ in a single gene, but in many of them. We shall consider in a later chapter the attempts to study developmental problems presented by single gene differences.

REFERENCES

Allen, E., et al. 1932. Sex and internal secretions. 951 pp. Williams and Wilkins Co., Baltimore.

Cockayne, E. A. 1938. The genetics of sex in Lepidoptera. Biol. Reviews, 13:107–132.

Crew, F. A. E. 1927. The genetics of sexuality in animals. 188 pp. Cambridge University Press.

Goldschmidt, R. 1934. Lymantria. Bibliogr. Genet., 11:1–186.

Schrader, F. 1928. The sex chromosomes. 194 pp. Gebr. Borntraeger, Berlin.

Schrader, F., and S. Hughes-Schrader. 1931. Haploidy in Metazoa. Quart. Review Biol., 6:411–438.

Wilson, E. B. 1925. The cell in development and heredity. 1,232 pp. The Macmillan Co., New York.

PROBLEMS

1. In the fowl, silver plumage is dominant to gold (brown) and is differentiated by a sex-linked gene. Give the results expected in F_1 and F_2 from a cross of silver female by gold male. Do the same for the reciprocal cross, gold female by silver male.

2. What results would be given by a sex-linked lethal in the fowl?

3. Strains of the moth Abraxas are known in which the Y chromosome is absent; in others a Y is regularly present. By means of diagrams indicate how the strain without a Y might be perpetuated.

4. Crew reported a case in the fowl in which a hen, which had functioned as a female, became transformed into a functional male as a result of development of a testis from the right gonad following destruction of the ovary by disease. What sex ratio would be expected in a mating between a normal female and such a transformed individual?

5. In some breeds of domestic sheep both sexes are horned, in others both sexes are hornless. The two conditions are differentiated by an autosomal pair of alleles. The horned condition is dominant in males but recessive in females. What results are expected in the F_1 and F_2 from the two matings between breeds, horned ewe by hornless ram and hornless ewe by horned ram?

6. Assuming two pure stocks of Habrobracon (a parasitic wasp in which the males develop from unfertilized eggs and are haploid), one homozygous for two independent recessive mutant genes and the other wild-type, give the results for F_1 and F_2 in the two matings between the two stocks, $a\ b\ ♀ \times +\ +\ ♂$ and $+\ +\ ♀ \times a\ b\ ♂$. (The sex ratio is, of course, not constant, since it depends on how many fertilized as compared with unfertilized eggs a female lays.) What would the expected results be with genes a and b linked with 10 per cent recombination?

7. In the liverwort Sphaerocarpos in which the gametophyte generation is the conspicuous one, the four spores that are produced from a single spore mother cell as a result of the two meiotic divisions remain together in a tetrad. They may be separated and individual gametophytes obtained from them. What types of spore tetrads, with respect to the genetic constitutions of the spores that make them up, would you expect from a sporophyte heterozygous for sex and for an autosomal gene?

8. Work out the types expected in Lepidoptera from matings of diploid females by triploid males.

9. What types of offspring are expected in Drosophila from the mating of a triploid female with two X chromosomes attached, by a normal male? Do not attempt to work out their expected relative frequencies—the distribution of chromosomes is not random. Indicate which of the types have attached-X chromosomes.

10. In the fowl, darkly pigmented connective tissue which can be identified by the external appearance of a bird (found in the Silky breed) is differentiated from normal by the simultaneous presence of an autosomal dominant gene and a sex-linked recessive gene. What results are expected in the F_1 and F_2 generations from reciprocal crosses between individuals that differ in both these loci?

11. It is extremely difficult to differentiate sexes in newly hatched chicks of the domestic fowl. In practice it is an advantage to be able to identify the sexes as soon as possible. Knowing that the sex-linked character barring can be classified immediately on hatching of chicks, outline a genetic method for separating males and females in young chicks.

CHAPTER XVII

OVERLAPPING PHENOTYPES, SELECTION, AND HYBRID VIGOR

Continuous Variability.—It is evident that many characters do not lend themselves to a classification into distinct groups. Such a character as height in man, for example, obviously varies continuously over a wide range; it is not possible to draw any sharp distinction between tall men and short ones. It so happens that many of the characters that are of practical importance in the breeding of cultivated plants and domestic animals show this kind of continuous variation. Examples are furnished by rate of egg production in fowls, butter-fat percentage in the milk of dairy cows, or yield in crop plants.

The inheritance of such differences as are met with in these cases cannot be studied directly by the methods discussed in earlier chapters. The very existence of such continuous variability was formerly looked upon as a serious objection to the whole gene theory. It has been found, however, that there is no contradiction here; special methods of study are needed, but the principles involved are the same as those that apply to discontinuous variations.

Interpretation of Continuous Variability.—We have already described (Chapter IV and elsewhere) a number of examples in which more than one pair of genes are concerned in the determination of a single character—*e.g.,* coat color in rodents, or eye-color in Drosophila. In these cases the number of gene-pairs segregating in a single experiment was small, and the phenotypes produced were sharply distinguishable; but neither of these relations is a necessary one.

It would be possible to build up a strain of Drosophila in which the eye-colors showed a practically continuous range from white to dark red, if one made the strain heterogeneous for a dozen or so known pairs of alleles. If, in addition, there were no strains available that were homogeneous and that differed from each other by single genes affecting eye-color, the analysis of the nature of the mixed strain would be very difficult.

This hypothetical situation is similar to the actual condition met with in many cases. Such examples usually arise as the result of long-continued selection—either natural or artificial. Many relatively slight mutant changes are present in heterozygous condition in most strains, and selection for extreme development of a given character is effective in piling up the more "favorable" alleles for all loci in which the strain is heterogeneous. In the hypothetical strain of Drosophila referred to above, selection would, in the course of time, enable one to produce true-breeding strains of either of the two extremes—red eyes or white ones—by accumulating, and gradually increasing the homozygosis of the alleles working in the desired direction; or, stated in words that are perhaps less misleading, by eliminating the alleles producing effects in the direction not desired.

Ear Length in Maize.—One method of studying continuously varying characters is through the crossing of two relatively constant strains that differ from each other. An example is illustrated in figure 101, from the work of Emerson and East on ear-length in maize. The short-eared parental race had ears ranging from 5 to 8 cm. in length, the long-eared one ranged from 13 to 21 cm. The F_1 plants had ears ranging from 9 to 15 cm.—*i.e.,* intermediate between the two parents, with a modal value of 12 to 13 cm. and a variability little (if any) greater than that of the parental strains. In F_2 the range was from 7 to 19 cm., the average lying close to that of the F_1 plants but the extremes coming well within the respective ranges of the two parental strains. The characteristic result here is the great variability of F_2 as compared with F_1.

Genetic Basis of Ear Length.—These results may be explained in terms of a number of pairs of alleles affecting ear-length. The parental strains differ in several such pairs, but each is relatively homogeneous. The F_1 plants are also relatively homogeneous, but are heterozygous for all pairs of genes in which the parental races differed. The intermediate nature of the F_1 may be interpreted in either of two ways: perhaps the short-eared strain had some dominant alleles for shortness and some recessive ones in other loci, the long-eared strain having the corresponding recessive alleles in the first kind of loci, the dominant ones in the second kind; or, perhaps complete dominance is absent, the vari-

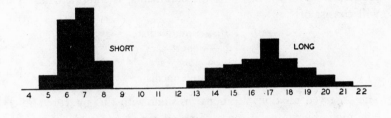

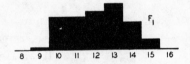

Fig. 101.—Frequency distributions for length of ears in maize. The horizontal axes represent centimeters, the vertical axes percentages of the four populations shown. (Based on data from Emerson and East.)

ous pairs of alleles giving intermediate heterozygotes. Presumably both kinds of relations occur, at different loci.

On the second interpretation, that of lack of dominance, a single pair of alleles is sufficient to account for the differences between the parental strains and the intermediate nature of the F_1; in that case, however, the F_2 should be made up of 1 homozygous short to 2 heterozygotes to 1 homozygous long. The addition of the two parental curves to twice the F_1 curve should, then, give a good approximation to the F_2 curve—an obvious disagreement with the facts. It follows that more than one pair of alleles is segregating in the F_1 plants.

If two pairs of genes are concerned, dominance may be present in both loci, in one, or in neither. In the former case we may designate the two pairs as *Aa, Bb,* the two parental strains may be supposed to be *AA bb* (short eared), and *aa BB* (long eared). *A* is then a dominant gene for short ears, *B* is a dominant for long ears. The F_1, having both dominants, is intermediate. F_2 will consist of

9 *A B* — intermediate
3 *A b* — short ears
3 *a B* — long ears
1 *a b* — intermediate

The observed relations are not consistent with this interpretation, but might be fitted by more complex ratios (27:9:9:3:9:3:3:1, etc.).

With two pairs of genes and no dominance—*i.e.,* an intermediate heterozygote—the number of classes is greater. The expected ratio here is 1:2:1:2:4:2:1:2:1. If we assume that the "long" alleles in each locus are equally effective, as are also the "short" ones, and that each heterozygote is exactly intermediate, the result becomes:

	Number of "long alleles"—*i.e.,* relative ear length				
	0	1	2	3	4
Proportion of plants	1/16	4/16	6/16	4/16	1/16

The assumptions made here are not likely to be exactly realized; they can be adjusted to give a reasonably good fit to the data if we assume partial dominance at one or both loci, or unequal effects of the two loci. Such adjustments are purely arbitrary; all that can safely be concluded from the data as given is that more than one pair of genes influencing ear-length is segregating.

The F_2 generation must include individuals like each of the parental strains, and in the case cited does include individuals like the modal classes. There can be no doubt that the failure to include the extreme parental classes is due to the relatively small size of the F_2 family (401 plants).

Number of Segregating Gene Pairs.—If the parental genotypes were capable of definite identification their frequency in F_2 would enable one to estimate the number of gene-pairs segregating—two pairs would give each parental type in $\frac{1}{16}$ of the F_2 individuals, three pairs in $\frac{1}{64}$, n pairs in $(\frac{1}{4})^n$. In the present case, however, as with most such characters, the variability of the parental strains is so great that their duplicates in F_2 cannot be positively identified.

Numerous attempts have been made to develop methods of estimating the number of gene-pairs involved in crosses such as this. All such attempts, however, must yield unsatisfactory results because of one circumstance: there is no way of determining the relative phenotypic differences conditioned by the various gene-pairs involved; these may range from fairly large ones down to those which are at the limit of the sensitivity of the measuring system used. The slight differences will play little part in the result unless they are numerous, and will be practically impossible to estimate numerically—yet there is no reason to expect them to constitute a class sharply distinct from the gene differences with larger effects. In practice, most attempts to solve this problem have started with the arbitrary and improbable assumption of equal effects of all gene differences concerned.

Determination of Genotypes.—There is one effective method of approach, however—viz., the study of the genotypical constitution, rather than the phenotypes only, of the F_2 individuals. If, in the case of ear-length discussed, it were possible to self each F_2 plant and rear an F_3 family from it, one should be able to determine what proportion repeated each of the parental curves, and in general to obtain more precise information regarding the composition of the F_2. The method has often been used, and has in many cases yielded satisfactory confirmation of the hypothesis of multiple genes.

Non-genetic Variability.—Another question arising in connection with continuous variation may be illustrated by reference to figure 101: if the two parental strains are genetically homogeneous, all individuals in each one having the same genotype, why does each strain vary in ear-length? Even in F_2, there should be a definite number of separate genotypes, and one might expect a series of distinct phenotypes—one corresponding to each genotype. It is, of course, obvious in everyday experience that such characters as size usually show no such relations; the number of classes obtainable is conditioned only by the smallness of the unit of measurement used. In the case of ear-length of maize it is easily observed that two ears borne on the same plant, and therefore alike in their genotypes, will almost always be of different lengths if a small enough unit of measurement is used. This gives the clue to the solution of the problem raised; a single genotype need not give always the same phenotype. This conclusion is obvious, but must be kept constantly in mind, even though it amounts only to stating that organisms may be influenced by external conditions—such as food supply, temperature, moisture, or mechanical accidents. In most cases, however, it is not permissible to attribute all the variability of the parental strains to environmental influences, since it is difficult to produce parental strains that can safely be assumed to be homozygous for all genes affecting the character concerned. We shall return to this matter later in the present chapter.

Transgressive Variability.—In the cross discussed above, the F_2 generation did not include individuals more extreme than either parental strain. This is a usual relation, but not a necessary one. It will be found whenever all the genes favoring one extreme are present in one parental strain, all those favoring the other in the second parental strain. However, exactly the same F_1 and F_2 might have been produced if some of the genes had been interchanged between the parental strains. In that case the parental strains themselves would have been less different from each other, and might well have included no individuals as extreme as some of those found in F_2. Such examples of transgressive variability in F_2 have been recorded.

Number of Offspring Like Parents.—If many genes are segregating, the most extreme classes in F_2 may be rare—1 in 4^n individuals, where n is the number of gene pairs segregating, and where dominance is incomplete. As will be pointed out below, if linked genes are concerned the most extreme combination possible with the given heterogeneity may not be attainable at all in F_2. In any case, it is likely to require so large an F_2 family, to be reasonably certain to obtain the most extreme types, that some other method must be adopted. The method used is that of selection.

If four pairs of genes are segregating, the F_1 may be represented as $A/a\ B/b\ C/c\ D/d$. If the complete homozygote, $a/a\ b/b\ c/c\ d/d$, is the required extreme type it is evident that it will occur only once in an average family of 256 F_2 individuals. Furthermore, because of non-genetic variability, one can not always be certain of identifying individuals of the desired type. There will, however, be eight times as many individuals that carry only one dominant allele, and any one of these will readily produce the desired extreme in F_3—in one fourth of its offspring if self-fertilization be carried out.

SELECTION

The same principle applies if, instead of an F_2 generation, one starts from a population of mixed and unknown origin, where the various alleles are present in various proportions and combinations. In such a case, selection of the desired type of individuals will increase the proportions of the alleles that affect the character in the desired direction in the next generation. It may take a number of successive generations of selection to obtain the maximum possible development of the character concerned, but this method will achieve the result more quickly than will random breeding; though in theory random breeding should in time produce all possible combinations of the genes present in the original population. In practice, the number of individuals required to get the desired result in this way is usually so fantastically large that no breeder considers the possibility.

Number of Bristles in Drosophila.—Figure 102 represents the results obtained in a typical selection experiment. MacDowell started with a strain of Drosophila in which occasional individuals had extra bristles. He selected, in each generation, those individuals with most extra bristles as parents for the next generation. In this particular case, though not in others we might have used as examples, brother-sister matings were always used. The curve shows that selection was effective in increasing the average number of extra bristles present, and further that the effectiveness was greatest in the earlier generations. In the later ones the curve fluctuates so much as to suggest that much of the variability re-

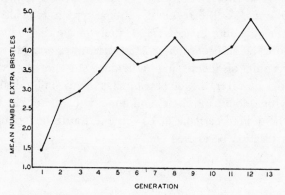

Fig. 102.—The effects of selection for extra bristles in Drosophila. (From data of MacDowell.)

maining after about five generations of inbreeding and selection was of environmental, rather than genetic, origin. The experiment was continued through 48 generations; the later generations do suggest that further progress was made after the 13 shown in the figure, though so little that improvement in the technique of culturing the animals remains a possible interpretation.

The slight increase in late generations of selection, just referred to, has been observed in a number of other selection experiments also; presumably in some, at least, of these instances it is due to the occurrence of new "favorable" mutations—which are to be expected occasionally.

Increasing Homogeneity.—If results such as these are due to sorting out and perpetuation of "favorable" genes already present in the population, the observed relations are the ones that are expected. There are several corollaries that were tested by Mac-Dowell. The degree of heterogeneity should constantly decrease in successive generations, since the effects of selection may be thought of as resulting merely in the elimination of genes that have effects in the "wrong" direction. Consequently, reversal of selection—*i.e.,* selection for low number of bristles—should be effective if started in early generations, ineffective if started in later ones. MacDowell found this to be true, as have other investigators using other material. The decrease in heterogeneity should also be detected in another way; so long as an appreciable part of the variability is genetic in origin, there should be a correlation between the numbers of bristles present in parents and in their respective offspring, if parents of diverse grades be tested. In fact, it was found that the parent-offspring correlations, at first rather high, gradually decreased to zero as the strain became more and more uniform in its genetic constitution.

INBREEDING AND HETEROZYGOSIS

Inbreeding itself, even without selection, leads to a decrease in heterogeneity. If a plant, heterozygous for a single pair of genes, is self-pollinated, the offspring will be in the proportion 1 *AA*: 2 *Aa* : 1 *aa*. That is, half of them will be homozygous. If selfing is again carried out, the homozygous individuals will give only homozygous offspring, the heterozygous ones will again give half their offspring homozygous. That is, on continued selfing the proportion of heterozygotes will be halved in each generation. A similar argument may be applied to an individual heterozygous for a large number of independently segregating pairs of genes; the number of pairs remaining heterozygous will be halved by each generation of selfing. Looking at the matter in another way, the chance that any one locus will remain heterozygous in a single individual chosen at random, will be halved at each generation. The same principle applies to any form of inbreeding. In no other case will the rate of increase of homozygosis be as rapid as on selfing, but brother-sister, double cousin, or other types of matings between relatives will lead to increases in homozygosis (Fig. 103). If such inbreeding is accompanied by selection, the rate of increase will, in general, be greater—since selection will lead to the mating together of individuals with like phenotypes and therefore more probably of similar genotypes.

Inbreeding and Selection.—This consideration leads to certain deductions concerning the efficiency of various methods of carry-

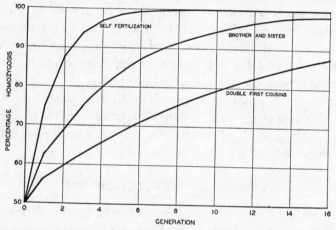

Fig. 103.—The effects of inbreeding on homozygosis. The vertical axis represents the percentage of independently segregating gene pairs for which the series was originally heterozygous but has become homozygous as a result of the type of inbreeding indicated. (After Wright.)

ing on selection experiments. If results are wanted quickly, it will obviously be desirable to inbreed closely, since this will lead to the most rapid elimination of the undesirable alleles. If, on the other hand, the aim is to get the most extreme possible individuals, then only moderate inbreeding should be practiced—since the object should be to avoid discarding any favorable alleles. In fact, wide outcrosses, to unrelated strains, followed by inbreeding and the rearing of large families, would appear to be the best method of obtaining the maximum genetic diversity from which to start selection.

In practical work, it is clear that many considerations will be involved in the choice of method. In horses, where each individual is expensive to rear, requires several years to reach sexual maturity, and produces relatively few offspring, the method of outcrossing will not be profitable. In the case of an annual plant, where the individual is of less consequence in terms of time, space, and expense, the efficient method of procedure is more likely to be that of first outcrossing.

LINKAGE

We have so far considered the relations concerned in continuously varying characters without reference to the occurrence of linkage. It is evident that genes lying in a single pair of chromosomes often affect the same character, and that populations may be expected to be heterogeneous with respect to such linked pairs of alleles. The exact analysis of such relations in practice is difficult; but it is certain that linkage is responsible for some of the difficulties encountered. One need only postulate the existence of a "favorable" allele in a single chromosome, with an "unfavorable" one located on each side of it, to realize that the most favorable combination of alleles possible may be an extremely difficult one to obtain in practice.

Linkage with Known Mutant Genes.—In a few cases it has been shown that some of the genes that affect characters of the type here considered are linked to other genes that have easily recognized effects. In theory it should be possible to employ this method, of using known mutant genes as markers, and thereby to obtain a complete analysis of the effect of the genes in each chromosome and each section of chromosome on any character. In practice the method is laborious, and in most organisms not enough known mutant genes are available. There exist, especially in Drosophila, enough partially analyzed cases of the sort to show that the theory is correct, even if its practical use is difficult.

One practical application of linkage may be suggested here, though its actual use involves so much labor as to limit its availability. If it can be demonstrated that certain desirable alleles are linked to known genes with easily classifiable effects, then the favorable genes may be recombined with other ones with much greater ease than before; the difficulty lies in establishing the linkage in the first instance.

MODIFIERS

Genes with small effects, of the kind under discussion in this chapter, are sometimes referred to as *modifiers* or *modifying genes*. The term is a loose one, sometimes being used only with the implication that the alleles in question do not result in sharply distinct phenotypes, and that there are several pairs of genes affecting the character in question. Another usage is more precisely indicated by the term *specific modifier,* meaning a gene without an easily recognizable phenotypic effect except in the presence of a particular constitution at some other locus. In this sense genes affecting eye-color in Drosophila (such as *v, pr, st*) are specific modifiers of *w+*—since they produce no recognizable phenotypic effect in white-eyed flies. In practice the term is most often used in connection with specific modifiers of mutant phenotypes.

HYBRID VIGOR

There is one continuously varying character that represents an interesting special case—viz., the general vigor or vitality of the individual. It has long been recognized that a decrease of vigor is likely to follow inbreeding. This relation was studied experimentally by Darwin; it was not a new idea even then. The modern interpretation is based largely on the work of Shull, of East, and of Jones. All these investigators studied the effects of inbreeding in maize.

Inbreeding and Vigor.—Maize plants are normally cross-pollinated. The pollen is produced in the tassel, and is scattered over the field by wind. It usually happens that most of the silks on a given plant are pollinated by grains from other plants; artificially, however, it is easy to insure self-pollination. When this is done, and the resulting seeds are sown, it is found that the plants produced are a rather poor lot. If these in turn are selfed, the second inbred generation is still less vigorous. The plants are commonly smaller and more difficult to grow, and bear smaller ears with fewer kernels on them, than is the case with the general population with which the experiment was begun. The general vigor decreases on inbreeding, and selection is powerless to prevent the deterioration—though it may make the difference between marked and less extreme deterioration. If selfing is con-

Fig. 104.—At right and left, plants of two inbred strains of maize. Center, the F_1 generation of a cross between these two strains. (Photograph from Dr. D. F. Jones.)

tinued for many successive generations, the decrease in vigor continues for some time, but the rate of decrease becomes less. Many strains become so weak that it is impossible to keep them, but the others finally reach a constant state; further inbreeding produces no effects.

Crossing Inbred Strains.—The separate inbred strains that have reached this constant state differ greatly among themselves phenotypically, but each is quite uniform within itself. That is to say, each is homozygous for most of its genes, but the different lines are homozygous for different genes. If two such inbred strains are crossed, the F_1 generation is again uniform; but its vigor is greatly increased. When a series of inbred strains are tested, it is found that the F_1 plants produced by some combinations are stronger and give markedly better yields of grain than do those of the original mixed population from which the inbred strains were derived.

Heterosis.—This phenomenon is known as *heterosis,* since the vigorous individuals are the ones with the maximum number of loci in the heterozygous state. As will be shown below, it is not probable that heterozygosis, *per se,* is responsible for the vigor; but there remains a striking parallelism. Inbreeding is, as shown above, bound to lead to increase in homozygosis, and we have just pointed out that selfing leads to a decrease in vigor in spite of selection for increased vigor. The general population is heterozygous for many genes, as shown by the diversity of the inbred strains that can be derived from it—and the vigor of the general population is good. It should happen that some of the inbred lines would differ in more genes than the average number that are heterozygous in the general population—and, as stated, some crosses of inbred lines do give F_1 plants that are more vigorous than the general population.

Basis of Heterosis.—These relations may be interpreted on the basis of these two known facts: (1) most mutations are deleterious, and (2) most mutations are recessive. Maize is evidently a plant in which many unfavorable recessive genes are present. Normal cross pollination prevents these occurring in homozygous form in sufficiently large numbers to make serious difficulties, though they do somewhat decrease the efficiency of the whole population and the yield of grain per acre. Inbreeding automatically produces plants homozygous for recessive genes. The regularity of the occurrence of the deterioration shows that every plant in the original field population is heterozygous for many such deleterious recessives. The failure of selection to counteract the unfavorable effect also shows that the number must be large. Since the number is large, it is to be supposed that several unfavorable alleles are likely to exist in some or all chromosomes. One chromosome will carry unfavorable alleles a, c, and d; its homolog in the original field plant may carry $a+$ $c+$, and $d+$, but also possess the unfavorable alleles b and e. That is, the constitution may be written $\dfrac{a + c \; d +}{+ \; b + + \; e}$. It may be practically impossible to recover a chromosome with all $+$ alleles, and accordingly all the homozygous types produced will have fewer loci at which there are $+$ alleles than did the parent. If several or all chromosomes are in conditions like this, it is evident that selection and inbreeding will never lead to the production of the best possible phenotype—one with all $+$ alleles represented.

Heterosis and Practical Breeding.—The principles involved here have been put to practical use in breeding high-yielding maize. A series of inbred strains has been developed, specific combinations of which give unusually vigorous and productive F_1 plants. In practice, however, this method suffers from one serious defect: the seed parent, from which must be harvested enough seed to sow many acres, is weak and relatively infertile. This difficulty has been surmounted by the use of four separate inbred strains—A, B, C, and D. A $\times$ B and C $\times$ D hybrids are obtained; each is vigorous and easily grown. They are then planted in alternate rows in a field, and the tassels are removed from all plants of one type—say the A $\times$ B hybrids. The effective pollen then comes from the C $\times$ D rows of plants, and all the seed harvested from the detasseled plants is of the ancestry (A $\times$ B) $\times$ (C $\times$ D). If the four original strains are well chosen—and by repeated trials highly favorable combinations can be found— the plants grown from this seed are very vigorous, and give high yields. There is more variability than in the F_1 plants, since both parents here are heterozygous; but the greater fertility of the seed- parent makes this method more efficient for large-scale production than is the use of F_1 plants directly.   A large and increasing portion of the commercial crop in the United States is now grown from seed of this kind.

Other examples of the use of heterosis in practical breeding may be cited, but are of a somewhat different nature, being based on the crossing of standard strains that are already in existence and are of practical use even without crossing.   The best-known of these examples is the mule—an F_1 hybrid whose vigor is proverbial.   Many fancy beef cattle are produced by the crossing of pure breeds—the Shorthorn-Angus combination being an especially noteworthy example.

Natural Self-pollination.—Some kinds of plants—*e.g.,* peas, beans, wheat, oats—normally reproduce by self-pollination. Crossing is possible, but rarely occurs except when purposely brought about by man.   In such forms the phenomenon of heterosis is less marked than in most naturally cross-bred forms.   The reason is clear: unfavorable recessive mutant genes become homozygous almost at once and are eliminated, while in cross-bred organisms the fact that they are recessive makes it possible for them to be carried along indefinitely in heterozygous condition.

Most such normally self-pollinating species are made up of individuals that are homozygous for nearly all their genes.   (The striking exception here is Oenothera, many species of which are self-pollinating balanced heterozygotes—see Chapter XI.)   One of the classical series of experiments in the history of genetics— that of Johanssen on beans—was based on this fact.   Johanssen started from a mixed lot of beans, and selected large and small seeds.   On growing plants from these, he found that the large seeds gave plants that, on the average, had large seeds; the small seeds gave plants with small seeds.   The seeds on each plant varied in size, but selection within such a family was wholly ineffective.   Evidently the original mixed lot of seeds included a series of different genotypes, each homozygous; the variability among the seeds of an individual plant was wholly phenotypical. The terms "genotype" and "phenotype" were first introduced by Johanssen in discussion of these experiments.

Value of Homozygous Strains.—Such homozygous strains are of value in the study of physiological problems. It is clearly desirable to have as uniform individuals as possible for physiological studies, in order that the results obtained with them may be uncomplicated by genetic differences in reactivity. It is for this reason that many studies in plant physiology are carried out with such plants as peas and oats. Numerous closely inbred strains of rodents (rats, mice, guinea-pigs) have been developed for similar purposes.

ASEXUAL REPRODUCTION

There is another method of producing individuals with like genotypes—viz., asexual reproduction. The asexually produced descendants of a single individual are members of a single *clone*. Such individuals are alike, since meiosis and fertilization are not concerned in their production, but they need not be homozygous. In fact, long-continued asexual reproduction regularly leads to the accumulation of unfavorable recessives. Since the heterozygotes are at no disadvantage and the homozygotes are not produced, new recessive mutations are eliminated only when an allelic recessive arises in the homologous chromosome.

Artificial Asexual Propagation.—Many cultivated plants are artificially propagated asexually, through the rooting of stem-cuttings (as in the common house "Geranium," Pelargonium) or tubers (as in potatoes), or through budding or grafting (as in many fruits—peaches, apples, etc.). In these cases seeds are produced, and are usually the chief means of reproduction in wild plants. In the cultivated strains they are the source of new varieties; but sexual reproduction, with its attendant recombination, leads to the production of seedlings, the great majority of which are inferior.

"Inferior" in this case is to be understood as meaning "less desirable from the point of view of the plant breeder," rather than "less efficient from the point of view of the apple." The seedlings are often vigorous and highly fertile, but bear apples of poor quality. Evidently quality is a complex character, and is conditioned by many genes; in a given asexually reproduced variety many of these are heterozygous, and sexual reproduction leads to much recombination—most of the products of which are "inferior."

Citrus Fruits.—A special type of asexual reproduction is found in citrus fruits (oranges, lemon, etc.). Here, as in most plants, the seeds are produced only if pollination occurs and is followed by fertilization and the production of a diploid embryo. In many forms of citrus, however, the purely maternal tissue in each seed also develops into one or more embryos, and these compete with the embryo resulting from fertilization. The result is that the sexually produced embryo, being the product of recombination for many gene pairs, usually has a genotype much less viable than that of the mother plant, and is crowded out. In consequence, most citrus seedlings are wholly maternal in genetic constitution.

TUMORS

Another special case of the action of multiple genes is illustrated by the work of Little, of Tyzzer, of Strong, and of others on transplantable tumors in mice. Using a particular tumor line, it is possible to find inbred strains of mice that are consistently immune—*i.e.*, no tumors develop on inoculation—and other inbred strains in which every individual is susceptible—*i.e.*, an implanted bit of tumorous tissue will develop. In the cases studied, when these two strains of mice were crossed, all of the F_1 mice were susceptible. In the F_2, however, the majority of the individuals were immune. In one case studied by Little, for example, there were 23 susceptible to 66 immune individuals. On crossing the same F_1 generation back to the immune parental strain, there were produced 21 susceptible to 208 immune mice. These results are most easily explained on the assumption that the two original strains differed in four independent pairs of alleles, susceptibility occurring only in individuals carrying a dominant allele in each of the loci.

The results here depend on the particular tumor used, as well as on the particular strains of mice. Cases have been recorded in which the number of gene pairs concerned was one, two, three, four, five and more. In case the same pair of strains is tested by the use of different tumors, it may happen that they differ in two genes conditioning the growth of one tumor, but by three conditioning the growth of another; and there may be some genes common to the growth requirements of the two tumors, others specific to each tumor.

Tissue Transplantation.—These results are to be related to those obtained by Loeb and Wright in studies on the transplantation of normal tissues in guinea-pigs. In ordinary random bred material it is possible to make successful grafts from one individual to another part of the same individual, but not usually between individuals. If, however, long inbred strains are used, tissue may be transplanted from one individual to another one of the same strain and it will grow practically as well as though implanted in the individual that produced it. That is, the specificity in reactions between host and implant are determined by genes, and within a closely inbred line the genes are so nearly alike from one individual to the next that there is no difference in specificity. Transplantation between different inbred lines is, as expected, unsuccessful. If two such lines are crossed, it is found that tissue from either of the component inbred strains will grow if implanted in the F_1; tissue from the F_1 will not grow in the inbred lines. That is, all the specific genes present in the implanted tissue must be present in the host to insure success, but the reverse is not true; specific genes present in the host need not be present in the implanted tissue.

Spontaneous Tumors.—The occurrence of spontaneous tumors presents a more complex problem than does the susceptibility to implanted ones discussed above. Most of the experimental work here has been done with mice. The results of Lynch and of others show that strains of mice differ in the frequency with which individuals belonging to them develop tumors, and crosses show that these differences are inherited as though multiple genes were involved. The analysis is difficult, because tumors do not develop at a specific age, and many mice die before one can be certain whether they would have developed tumors or not. The incidence is also greatly affected by environmental conditions, and the situation is complicated by the existence of many kinds of tumors—some of which at least show different relative frequencies in different strains. All that can be said briefly is that strains differ genetically in the probability of the development of a spontaneous tumor.

It is probable that this conclusion may be applied to man as well. Analysis of pedigrees gives some support to such a conclusion; but most geneticists will be inclined to rely on the *a priori* argument that anything so complex and so dependent on many factors must, of necessity, be influenced by genetic differences.

IDENTICAL TWINS

This same *a priori* argument may be applied to the inheritance of mental characteristics in man. Here again there is some evidence from the analysis of pedigrees that goes to show the inheritance of some forms of insanity and feeble-mindedness. The number of types here is so great, however, and the influence of environmental differences is so difficult to evaluate, that extreme caution is indicated in drawing conclusions concerning the inheritance of mental characteristics.

There is one type of evidence concerning such characters that is particularly useful. This concerns identical twins, which may be looked upon as representing a special case of asexual reproduction, already referred to as a method of obtaining genetically identical individuals. Identical twins are produced in about 1 in 400 human births. Both members of such a pair develop from a single fertilized egg, and they must therefore be supposed to be genotypically identical. It is a matter of common knowledge that identical twins are in fact closely similar in appearance and in mental make-up. More detailed studies have shown a whole series of curious resemblances—in finger-prints, tooth anomalies, disease incidence, etc. Such pairs also offer the best hope of providing critical evidence on the old question of the relative rôles of heredity and environment in the determination of mental characters in man. The two members of a pair of identical twins are alike in their genes; what is required is to find pairs the members of which have been reared apart from infancy. Muller and later Newman and others have studied some 20 such separated pairs, and have shown striking similarities to exist in their personal characteristics, as indicated by comparative scores in various kinds of "intelligence tests" and by other less objective studies. In general, identical twins reared apart seem to resemble each other, mentally as well as physically, more than do non-identical twins (*i.e.*, twins developed from two separately fertilized eggs). Further studies of this sort will furnish a really sound basis for the analysis of the hereditary basis of human personality.

REFERENCES

Babcock, E. B., and R. E. Clausen. 1927. Genetics in relation to agriculture. 673 pp. McGraw-Hill Co., New York.

Bittner, J. J. 1938. The genetics of cancer in mice. Quart. Review Biol., 13:51–64.

East, E. M. 1936. Heterosis. Genetics, 21:375–397.

East, E. M., and D. F. Jones. 1919. Inbreeding and outbreeding. 285 pp. J. B. Lippincott Co., Philadelphia.

Newman, H. H., F. N. Freeman, and K. J. Holzinger. 1937. Twins: A study of heredity and environment. 369 pp. Univ. Chicago Press, Chicago.

Sinnott, E. W., and L. C. Dunn. 1935. The effect of genes on the development of size and form. Biol. Reviews, 10:123–151.

PROBLEMS

1. How many different genotypes are possible in a population heterogeneous for a single pair of genes? How many if two pairs are segregating? Three? Ten? What is the general rule?

2. Given a plant of a species in which either selfing or cross-pollination is possible, of the constitution $\dfrac{+\quad b\quad +}{a\quad +\quad c}$: Suppose the crossover values to be: *a-b*—10 per cent, *a-c*—15 per cent, *b-c*— 5 per cent (no double crossovers); how would you obtain a strain homozygous for *a b c?*

3. In the example of inheritance of ear-length in maize given in the text, the actual data recorded by Emerson and East were as follows (in percentages):

Ear length (cm.)	Short parental strain	Long parental strain	F_1	F_2
5	7.0			
6	36.9			
7	42.0	....	....	0.3
8	14.0	....	....	2.5
9	....	....	1.4	4.7
10	....	....	17.4	6.5
11	....	....	17.4	11.7
12	....	....	20.3	18.2
13	....	3.0	24.6	17.0
14	....	10.9	13.1	17.0
15	....	11.9	5.8	9.7
16	....	14.8	....	6.2
17	....	25.8	....	3.7
18	....	14.8	....	2.2
19	....	9.9	....	0.3
20	....	6.9		
21	....	2.0		

On the assumption that there is a single pair of genes differentiating the two parental strains, show what distribution would be expected in F_2, on the basis of the parental and F_1 measurements. How does this expectation differ from the F_2 distribution actually obtained?

4. Selection for a character, such as milk production in cattle, that is phenotypically expressed only in one sex, requires the development of some system of selecting the breeding animals of the other sex. What basis would you suggest for selecting bulls to use in an effort to increase the milk production of a herd?

5. The brown scarlet double recessive in Drosophila has white eyes. Given a brown scarlet stock, and another one with the sex-linked recessive white, how would you make up a triple recessive strain (*w bw st*)?

6. The facet counts on bar and wild-type Drosophila described in Chapter XIV were collected to show the effects of differences at this locus alone. To do this, it is necessary to eliminate differences in genes at other loci (in the same chromosome pair or in other ones) that may affect facet number. Given an inbred wild-type strain, assumed to be homozygous for genes influencing facet-number, and given a bar strain with different genes in

other loci, how would you obtain a strain carrying the gene B but otherwise with nearly the same genes as those present in the inbred wild-type stock? (This result can be most efficiently obtained through the use of mutant genes in other loci as "markers." The process requires several generations.)

7. Two strains of mice differ in their susceptibility to two implanted tumors, one strain being susceptible to both tumors, the other to neither. For tumor number one the difference is dependent on two pairs of genes; for tumor number two it is dependent on these same two pairs of genes and one additional pair. Each tumor will develop only in mice carrying a dominant allele in each of the loci influencing it. The two strains of mice are crossed. What is the reaction of the F_1 mice to each tumor? What proportion of the F_2 mice will be susceptible to both tumors, to one and not the other, and to neither?

8. Two strains of Drosophila differ in a measurable character that shows continuous variability. How would you find out if some of the gene differences are in the X chromosomes? If one makes use of a standard test strain, carrying known dominant genes in each of the autosomes, it is possible also to test the two strains for differences in each autosome. Assuming that curly (Cy) is used for the second chromosome and dichaete (D) for the third (the fourth being neglected), outline the necessary crosses. (Remember that there is no crossing over in the Drosophila male.)

9. The standard ornamental varieties of roses are propagated asexually, through cuttings and through budding. They are heterozygous for many genes. Recently a technique has been developed which greatly reduces the time required for sexual reproduction, from pollination to flowering of the seedling. What practical advantage of this technique can you suggest?

CHAPTER XVIII

HETEROGENEOUS POPULATIONS

THE most usual method of approach to genetic problems is through the making of controlled matings, usually involving distinct strains, each known or presumed to be uniform within itself for the genes studied. In many cases, however, this method is not available. In man, for example, controlled matings are not possible; in horses they involve too much time and expense to be practicable. In still other cases, it is desirable to make some analysis of existing populations without being able to obtain any data at all on pedigrees of individuals. This latter situation is especially frequent in studies concerning wild populations. The development (by Haldane, Fisher, Wright, and others) of methods for studying such cases has also led to great advances in our theoretical knowledge of the behavior of genetically mixed populations, and of their reactions to mutation, selection, and inbreeding.

POPULATIONS IN EQUILIBRIUM

The simplest case is that of a population in which two alleles— A and a—are present. Let it be assumed that: (1) the population is indefinitely large, so that sampling errors may be disregarded; (2) mating is at random—*i.e.*, any two individuals of opposite sex are as likely to mate as are any other two; (3) both A and a are stable—*i.e.*, mutations may be disregarded; (4) the three genotypes AA, Aa, and aa are equally viable and fertile. We may suppose that the relative frequencies of the two genes are as p is to q, where p + q = 1—*e.g.*, if A is three times as frequent as a, then p = 0.75 and q = 0.25. Under the stated conditions, it is then possible to determine the proportions in which the three genotypes (AA, Aa, aa) will occur in the population. The gametes in any one generation will have constitutions proportional to the *gene frequencies*—*i.e.*, those carrying A will be to those carrying *a* as p is to q. A checkerboard diagram may therefore be constructed, as shown:

		Sperms	
		p *A*	q *a*
Eggs	p *A*	p² *AA*	pq *Aa*
	q *a*	pq *Aa*	q² *aa*

The three genotypes will be present in the proportions $p^2\ AA$: $2\ pq\ Aa$: $q^2\ aa$. This is the fundamental formula, an understanding of which is necessary in population studies. It represents a stable condition, which will persist unaltered so long as the conditions enumerated above are fulfilled. If the population represents a newly made mixture, the frequencies indicated will be present by the first generation after the original mixing occurs.

An example will show the usefulness of the formula. As stated in Chapter III, there exists a difference in ability of individual people to taste phenyl thiourea. Approximately 70 per cent find it extremely bitter; the remaining 30 per cent find it without taste. Preliminary analysis of a few pedigrees showed that two taster parents might have children who were non-tasters; but no cases were found in which two non-taster parents had children able to taste the substance (Snyder). This result suggested the hypothesis of a pair of genes, t^+ and t, the faculty of tasting phenyl thiourea being due to a single dominant gene. On this assumption, the gene-frequencies may be calculated:

$$p^2 + 2\ pq = 0.70$$
$$q^2 = 0.30$$
$$\therefore p = 0.45,\ q = 0.55,\ p^2 = 0.203,\ 2\ pq = 0.495$$

Given these values, it is possible to estimate the frequencies to be expected from two taster parents, or from parents of whom one is a taster, the other is not. These expectations follow from the calculated frequencies of $++$ (20.3 per cent) and $+t$ (49.5 per cent). For example, since $++$ to $+t$ = 2:5 (approximately), a series of taster parents will give, on the average, $9 +$ to $5\ t$ gametes, or about 36 per cent t. The observed results from taster $\times$ non-taster were 242 tasters and 139 non-tasters, or 39.5 per cent non-tasters. In this way the hypothesis of a single gene difference may be adequately tested without the accumulation of a large number of pedigrees that run through several generations.

Effect of Sex-linkage.—The formula requires a modification when the gene-pair concerned is sex-linked. In this case, the individuals of the heterozygous sex are haploid for the chromosome concerned, and the proportions in which their phenotypes occur give the gene frequencies directly; the usual formula applies to the homozygous sex. Red-green color-blindness occurs in about 8 per cent of human males; this then is the gene-frequency. The percentage of females that are color-blind is expected to be about 0.64 since this is the square of the gene-frequency. The available figures give frequencies for color-blind women that are definitely higher than this, for reasons that are unknown. In this case analysis of pedigrees was first used to show that color-blindness is a sex-linked recessive; but the gene-frequency method is available for use as a means of detecting sex-linked characters. If it be found that, in a mixed population, the frequency of a given character in the homozygous sex is the square of the frequency in the heterozygous sex, a sex-linked gene is to be suspected. A possible check on this interpretation is to see if the same algebraic relation holds in a different population in which the absolute numbers—*i.e.*, the gene frequencies—are different. The next example illustrates the use of this principle.

Blood Groups in Man.—The inheritance of the four blood-groups in man has been described in Chapter XII. The four types (O, A, B, AB) represent the interaction of three alleles, *O, A, B,* of which *O* is recessive and the other two produce their characteristic effects even when both are present. As a matter of fact, the original interpretation was in terms of two independent pairs of alleles, *Aa* and *Bb*, the O type being *aa bb*, the others of the various constitutions possible on the assumption that both *A* and *B* are dominant. It was pointed out by Bernstein and by Snyder that these two hypotheses lead to different equilibrium frequencies for the four phenotypes, as shown in the following example based on 8662 German cases recorded by Gundel:

Group	O	A	B	AB
Equilibrium,				
Multiple alleles	37.3	43.5	13.2	5.8
Two gene pairs	41.1	40.0	9.6	9.4
Observed	37.3	43.7	13.4	5.7

Evidently the hypothesis of three alleles gives a better fit than does the older view. It happens that the frequencies of the four types vary from one race to another. The American Indians are mostly O, B is most frequent in Asia, etc. It is found in every case that the hypothesis of multiple alleles gives the better agreement between calculated and observed phenotype frequencies, however varied the gene frequencies concerned. This conclusion has been verified by examination of individual pedigrees; the simplest diagnostic relation here is that, on the multiple-allele hypothesis, O and AB can never be related as parent and offspring; while on the hypothesis of two pairs of genes such pedigrees should be rather frequent. No clear cases of such a relationship have been found.

Limitations of the Formula.—The several conditions laid down above are necessary in order that the formula $p^2 + 2pq + q^2$ should be applicable. None of these conditions is necessarily fulfilled in every example. It is, however, possible to determine mathematically the consequences of introducing deviations from these requirements. The details of this mathematical development are beyond the scope of this book, but some of the conclusions may be indicated.

SELECTION

Selection, either natural or artificial, will upset the equilibrium. Referring back to the stated requirements, this is equivalent to saying that the three genotypes *AA, Aa,* and *aa* are not equally viable and fertile. It is evident that, if a population exists in the equilibrium proportions, but the *aa* individuals are wholly or relatively sterile, there will be fewer *a* gametes contributing to the composition of the next generation, and the composition of the population will change. The effects produced will depend on the intensity of selection, and on whether the selected phenotype is conditioned by a dominant or a recessive gene. It is obvious that a dominant gene, being subject to selection in every zygote that carries it, will change its frequency in response to selection much more rapidly than will a recessive gene, which is selected only when present in homozygous form. The difference in effectiveness will also depend on the gene-frequency itself, since the higher is the frequency of a recessive gene the more numerous are the homozygous individuals in which it will occur.

Selection and Rate of Change.—Haldane has calculated the number of generations required to bring about certain specified changes in phenotype frequencies, in the case where selection is of such an intensity that the less favored phenotype has a disadvantage of 1 in 1,000; *i.e.,* on the average contributes 999 gametes to the formation of the next generation when a like number of the more favored phenotype contributes 1,000.

TABLE 2

Character selected	Generations required to change phenotype frequencies as indicated			
	From .001 to 1%	From 1 to 50%	From 50 to 99%	From 99 to 99.999%
autosomal recessive	309,780	11,664	4,819	6,920
autosomal dominant	6,920	4,819	11,664	309,780
recessive, only one sex selected	619,560	23,328	9,638	13,841
dominant, only one sex selected	13,841	9,638	23,328	619,560
sex-linked recessive	10,106	5,593	4,668	6,916
sex-linked dominant	6,916	4,668	5,593	10,106

(The frequencies shown are related to gene-frequencies thus: for recessives, frequency shown = q^2; for dominants, $p^2 + 2pq$. The last two lines give the number of generations required to change the homozygous sex.)

If a selection 10 times as intense (*i.e.,* a disadvantage of 1 in 100) be assumed, the numbers of generations shown must be divided by ten in order to determine the number of generations required for a specified change.

INBREEDING

The assumption of an indefinitely large population whose members mate at random is probably never actually fulfilled. It is a convenient mathematical assumption; but it is evident at once that mere spatial propinquity will insure that, if the population is as large as stipulated, there will be some degree of selective mating, in the sense that fairly closely related specimens produced in a single area will be more likely to mate with each other than with more distantly related individuals produced at a remote point in the range. This question of the existence of relatively distinct sub-populations has been investigated mathematically by Wright. He has come to the conclusion that the condition optimum for adaptive changes in gene frequencies is that in which a large general population is split up into a series of small semi-independent sub-groups. These groups are small enough to allow random sampling to produce chance deviations in gene-frequencies; these new combinations are then "tried out," and any that are improvements spread to the rest of the area of the species; the unsuccessful tries are ultimately replaced by immigrants from the rest of the area occupied by the species.

MUTATION

The condition was laid down that the equilibrium formula is valid only when mutations are absent. Considering the case of two alleles, *A* and *a,* with gene frequencies p and q, there are two possible mutant changes: $A \rightarrow a$ and $a \rightarrow A$. We may represent the frequency of the first by the symbol u, of the second by v. The rate of change of p will then be represented by Δp, it being understood that the time interval is the same in estimating u, v, and Δp (commonly a single generation). It can be seen that $\Delta p = q v - p u$. When $\Delta p = 0$, the mutation pressures will evidently be at equilibrium—*i.e.*, there will be no further change in the constitution of the population. The condition of equilibrium then is $q\,v = p\,u$.

So long as neither u nor v reaches zero, it will be impossible, by selection, to obtain a completely homozygous population. In practice, however, the values of u and v are usually so low that populations homogeneous with respect to specific loci are frequently encountered.

If all loci are taken into consideration the situation is quite otherwise. Mutations at any one locus are usually too rare to affect calculations based on specific limited populations; but if all loci are taken into account one may be certain that diverse alleles will be found in any large cross-fertilized population. We have already described an example of this in maize, and many others are known. In wild populations of Drosophila unfavorable genes are present in the autosomes with surprisingly high frequencies— about 10 per cent of the second chromosomes of wild individuals of D. melanogaster carry lethal genes (Dubinin and co-workers), and the frequency is even higher in the third chromosome of D. pseudoobscura.

Mutations are the raw material for evolution. Selection and inbreeding cannot make any changes except in the frequencies of genes originally produced by mutation, which is thus the primary factor of evolution. Yet most mutations are deleterious; evolutionary progress is therefore dependent on a process that is usually injurious to the population in which it occurs.

REFERENCES

Dobzhansky, T. 1937. Genetics and the origin of species. 364 pp. Columbia University Press, New York.

Fisher, R. A. 1930. The genetical theory of natural selection. 272 pp. Clarendon Press, Oxford.

Haldane, J. B. S. 1932. The causes of evolution. 234 pp. Harper and Bros., New York.

PROBLEMS

1. Suppose you found two distinguishable types in a wild population of some organism, in these frequencies:

	Type 1	Type 2
Females	99%	1%
Males	90%	10%

   What would you conclude? If, in another organism, you found these same frequencies with the sexes reversed (*i.e.*, Type 2 in 10 per cent of the females and 1 per cent of the males), what would you conclude?

2. In Drosophila repleta there is a sex-linked recessive, "light," that affects the body color. It occurs in wild populations. One series, collected in New York City, included 6 lights among 172 females, and 33 lights among 224 males. What gene-frequency may be deduced from the males? Assuming stable equilibrium conditions, how many light females would have been expected in a sample of 172?

3. There are two antigens in human blood, known as M and N. No antibodies for these are present in human blood, so no agglutination occurs, and no clinical complications arise, in transfusion. Antibodies are produced in rabbits into which appropriate human blood is injected, and this permits classification of human subjects. The inheritance is independent of that of the A, B, and O types. One series of data (Wiener, Rothberg, and Fox) gave the following proportions of the three known types in a population:

   M, 29.3 per cent; MN (*i.e.*, both antigens present), 50.5 per cent; N, 20.2 per cent.

   What is the indicated genetic relationship of these types? Pedigree studies have verified the conclusion that is indicated; what are the expectations from the six possible kinds of matings?

4. Two separate populations are in equilibrium for the same pair of alleles. In one of them p = 0.6, in the other p = 0.2. If a random sample of females from one population is crossed to a random sample of males of the other, what is the result in F_1? (This may be solved most easily by the use of a checkerboard diagram.)

5. Suppose a pair of alleles to exist, with the following mutation frequencies:

$$+ \longrightarrow a = 0.001$$
$$a \longrightarrow + = 0.005$$

What will be the equilibrium frequencies for these alleles? What phenotype frequencies (with complete dominance of $+$) will be found in a population that has reached this equilibrium?

6. White seedlings in maize are effectively lethal, never growing into fertile plants. In spite of the resulting selection against the recessive genes responsible for them, commercial varieties of maize are heterogeneous for many such lethal genes. Why?

7. Using a checkerboard diagram, work out the equilibrium formula for populations heterogeneous for the three alleles concerned in the A, B, O blood groups in man.

8. In Primula acaule, the English wild primrose, two types of plants occur: one with short styles and long stamens, the other with long styles and short stamens. The difference is due to a single pair of alleles, short style being dominant. As shown long ago by Darwin, pollination of short times short, or of long times long is much less often successful than long times short or short times long. This relation appears to depend on a difference in rate of pollen-tube growth (compare self-sterility, Chapter XII). Examination of populations of wild plants shows the two phenotypes to be present in very nearly equal numbers in every large series studied. What frequencies of genotypes are present?

CHAPTER XIX

POLYPLOIDY

OFTEN, particularly in plants, more than two complete sets of chromosomes are present—a condition known as *polyploidy.* Organisms in which there are more than two sets of chromosomes all of which are derived from the same species are called *auto-polyploids,* while those in which the different sets were originally derived from two or more different species are known as *allo-polyploids.* This classification breaks down in certain cases in which the situation is more or less intermediate, but in general the distinction is convenient.

B-type Chromosomes in Maize.—It should be pointed out that organisms may have three or even four times the haploid number of chromosomes without being polyploid. One example of this situation, to be discussed in more detail later (Fig. 110), is found in the genus Drosophila. Here a species with twice the number of chromosomes of another may apparently have essentially the same chromatin organized into twice as many units, *i.e.,* a single chromosome of one species may be the equivalent of two in the other. The species with the larger number is not a polyploid. Another example is found in maize. Here supernumerary chromosomes are found in some strains. These extra units, known as "B-type" chromosomes, are made up largely of hetero-chromatin. They are genetically relatively inert, carrying no known genes. By selection they can be accumulated without important phenotypic effects. Randolph has obtained plants with as many as 10 B-type chromosomes (Fig. 105). Plants of these types are not triploids, for the reason that they do not have more than two *complete sets* of chromosomes—*i.e.,* of euchromatic material.

Origin of Polyploids.—Polyploids can arise in several ways, some of which will be discussed in detail later in this chapter.

Spores or gametes may be produced by anomalous meiotic divisions in which there is no reduction in chromosome number. Somatic doubling of chromosomes occasionally occurs in both plants and animals, and this may lead to the formation of diploid gametes. In mosses and ferns it has been possible experimentally to cause gametophytes to regenerate directly from a sporophytic tissue. Such gametophytes are, of course, diploid, and produce diploid gametes. The process can be repeated several times, each time doubling the chromosome number. Finally, the participa-

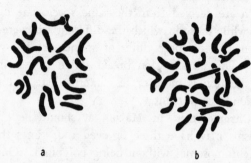

a   b

Fig. 105.—Somatic chromosomes of maize. *a*—normal figure, with 20 chromosomes (10 pairs). *b*—figure with 28 chromosomes, of which 8 are "B-type." These are among the smaller chromosomes, and are mostly to be found near the periphery of the metaphase plate. (After Randolph.)

tion of more than two nuclei in fertilization, which regularly occurs in the initiation of endosperm tissue in seed plants, can result in the addition of extra whole sets of chromosomes.

AUTOPOLYPLOIDS

Meiosis in Autotriploids.—Autotriploids have three complete sets of chromosomes. Every homolog is represented three times, just as is one specific chromosome in a trisomic individual. Each group of three homologs goes through meiosis in essentially the same way as does the single group of three in a trisomic individual. In maize, for example, in which the haploid number of chromosomes is ten, there are from seven to ten trivalents formed in the triploid. The orientation, with respect to one another, of these trivalents at metaphase is approximately a random one. The spores formed may have any chromosome number from 10 to 20, *i.e.*, from the haploid to the diploid number. In both male and female gametophytes the great majority of the unbalanced types are inviable, and a triploid plant is very nearly completely sterile. Haploid and diploid spores are expected to be produced with a low frequency—1 in 2^{10}, or 1 in 1,024, of each with random distribution and with no lagging. Both of these types are produced and are functional in both kinds of gametophytes. A few n + 1 pollen grains may function. In the female gametophytes many more combinations are functional than in the pollen.

Triploid-diploid Crosses.—If a cross is made between triploid and diploid maize plants, the few viable seeds obtained will give rise to a variety of types with respect to chromosome number. In fact, every number from 20 to 30 (diploid to triploid) may be found among the progeny. Trisomic types are relatively frequent, and it is from such crosses of triploid and diploid plants that most of the trisomic types in maize have been obtained. The reciprocal cross, diploid $\times$ triploid, gives mainly diploids, trisomic types and triploids.

The offspring of the cross, triploid $\times$ diploid, are of interest in connection with the genic balance concept previously mentioned. With regard to vigor the diploid and triploid plants are normal and those with intermediate numbers are less vigorous. Furthermore, those closest to the 2n and 3n conditions (*e.g.*, 2n + 1 and 3n − 1) show the least decline in vigor and those furthest removed (*e.g.*, 2n + 5) show the greatest decline. The correlation between chromosome number and phenotypic modification is not as simple as the above general statement might imply; nevertheless, the relations are in general agreement with those expected on the balance theory, if one remembers that the different chromosomes are qualitatively unlike.

Autotetraploids.—An individual with four complete chromosome sets is known as an *autotetraploid*. Autotetraploids usually show a high frequency of quadrivalent formation at meiosis. The leptotene chromosomes pair by twos, with more or less frequent changes of partners along the lengths of a group of homologs (Fig. 106). At least three chiasmata, appropriately related to

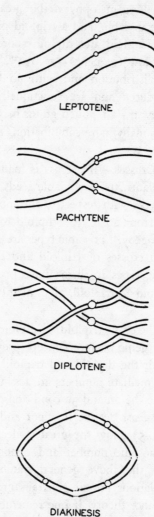

LEPTOTENE

PACHYTENE

DIPLOTENE

DIAKINESIS

Fig. 106.—Meiotic prophases in an autotetraploid, showing conjugation of four homologous chromosomes. The diakinesis configuration shown results from complete terminalization.

one another, are necessary for the formation of a metaphase quadrivalent. Frequently four homologs are associated as two bivalents or, less frequently, as a trivalent and a univalent. In the tomato, for example, there are 12 chromosomes in the haploid; and in the microsporocytes of the autotetraploid form there are from 7 to 11 quadrivalents and from 10 to 2 bivalents (Lesley).

Since quadrivalents usually show a 2–2 distribution at meiosis, autotetraploids are usually relatively fertile. The spores or gametes formed with regular distribution are, of course, diploid. The frequencies of 3–1 distributions for quadrivalents are sufficient, however, to result in rather numerous deviations in chromosome numbers in the offspring of autotetraploids.

Higher Polyploids.—Still higher autopolyploids are known. The wild tulip species, Tulipa clusiana, is an autopentaploid (5n). In this case there is an odd number of chromosomes of each kind and consequently meiosis results in irregular distribution. The species maintains itself in nature, without recourse to meiosis and fertilization, by resorting to vegetative reproduction. Crepis biennis, hawk's beard, appears to be an auto-octoploid. Other examples are known in nature and under cultivation.

Segregation in Autopolyploids.—Segregation of genes heterozygous in autopolyploids is, of course, different in detail from that in diploids. In autotriploids it is essentially similar to that already discussed in trisomic types. Because of the sterility of autotriploids and the complications resulting from the elimination of many types of gametes, genetic studies are particularly difficult to carry out. Genetic studies of triploid females of Drosophila have shown that crossing over occurs after chromosome division (*i.e.,* at a stage at which six chromatids are present), as it does in diploids, and have given information bearing directly on the pairing behavior and on the relation of crossing over to chromosome distribution. These questions are, however, too involved to justify a consideration of them here.

Segregation in autotetraploids occurs in a predictable manner. In the autotetraploid form of Primula sinensis, it has been shown that plants of the constitution $+/+/a/a$ give the gametic types $+/+$, $+/a$, and a/a in approximately the ratio 1:4:1 expected with random two-by-two assortment of four entire chromosomes.

The genotype $+/a/a/a$ gives approximately equal numbers of $+/a$ and a/a gametes. The above ratios hold only for genes which give no crossing over between the centromere and the gene concerned. For loci that do give such crossing over they do not hold strictly, but are roughly approximated. That such crossing over does occur is demonstrated by the fact that a/a gametes can be obtained from individuals of the constitution $+/+/+/a$. The exact genetic ratios expected in tetraploids depend on the frequency with which crossing over occurs between the centromere and the locus of the gene being studied. Since this has so far been determined only from the observed ratios themselves, and since the relation between the two is somewhat involved, we shall not consider it further here.

ORIGIN OF POLYPLOIDS

It occasionally happens in most plants and animals that, through irregularities in meiosis, non-reduced spores or gametes are produced. As a result of the formation of diploid gametes it is possible for either triploids or tetraploids to arise, depending on whether one or both gametes are diploid. In most normal diploids the frequency of formation of diploid gametes by such irregularities in meiosis is so low that the probability of obtaining tetraploids by the chance meeting of two such gametes is practically zero.

Through breakdown of the synchronism between chromosome multiplication and division of cytoplasm during mitosis, it is possible for diploid cells to give rise to tetraploid cells. The descendants of such cells give rise to islands or sectors of tetraploid cells. If such sectors happen to be in the germ line, diploid gametes or spores are formed as the result of normal meiotic divisions. The results are, of course, essentially the same as those following non-reduction at meiosis. The frequency of such somatic doubling of chromosomes can be greatly increased artificially. Randolph has been successful in inducing tetraploidy in maize by short sub-lethal heat treatments applied soon after pollination. This may double the chromosome number during the first division of the fertilized egg, and thus give rise to an entire tetraploid plant. Self-pollination then establishes a tetraploid strain. Crosses between tetraploids and diploids give triploids in F_1.

Decapitation of such plants as tomatoes leads to the production of new shoots from adventitious buds in the callus tissue formed at the cut surface. For some unknown reason such shoots are frequently tetraploid. Either sexual or vegetative reproduction from such shoots can be made use of in establishing a strain with the double number of chromosomes. It has been found recently that treatment with the alkaloid colchicine greatly increases the frequency of somatic doubling of chromosomes in a number of plants. Decapitation followed by treatment with indole-3-acetic acid (a plant growth hormone) has been found to lead to a similar end result in tobacco.

HAPLOIDS

In connection with polyploids, it is convenient to consider organisms with a single complete set of chromosomes. These are known as haploids. The gametophyte phase of plants is, of course, regularly haploid; and in most of the lower plants, Thallophytes and Bryophytes, this is the conspicuous phase. Haploid sporophytes are known in many species of higher plants as exceptional individuals. Incidentally, the fact that the sporophyte can be haploid and the gametophyte diploid shows that the normal haploid-diploid relation between gametophyte and sporophyte is not a causal one. Exceptional haploid individuals may occur in many animals. A number of species of animals—all of the Hymenoptera (bees and wasps), several other insects, and a few other forms—regularly have haploid males (see further discussion in Chapter XVI).

Meiosis in Haploids.—In sporadic haploids, meiosis is irregular as a result of the absence of chromosome pairing. Since each chromosome is represented only once, only univalents are present at the first meiotic division. These are distributed to the two poles more or less at random. Most of the spores or gametes will be deficient in several chromosomes. With random distribution of the univalents in a maize haploid, only one ten-chromo-some spore in 1,024 divisions is expected, *i.e.*, one in 2^{10}. As a consequence of these and other irregularities in chromosome behavior, haploids are characteristically highly sterile. An inter-esting use can be made of monoecious haploid plants in obtaining homozygous types. Obviously, if a haploid produces offspring by self-fertilization, such offspring will be completely homo-zygous; the two sets of homologs are immediate descendants of a single set.

In species of animals in which the males are regularly haploid, meiosis is modified in such a way that sperms are produced with-out a reduction in the chromosome number. Thus, in many of the Hymenoptera spermatocytes undergo a first meiotic division in which one of the products is an abortive cell with no chromo-somes. The daughter cell containing all the chromosomes then undergoes a regular second meiotic division, as a result of which two haploid spermatids are formed (page 249).

Haploid Parthenogenesis.—Haploid males in Hymenoptera regularly arise as the result of the direct development of an egg without fertilization, a process known as *haploid parthenogenesis*. In this order the regular rule is that fertilized eggs develop into diploid females and unfertilized eggs into haploid males (see page 248). Haploid parthenogenesis may occasionally occur spon-taneously in species belonging to other groups. In many species it may be induced artificially. Haploid parthenogenesis may be followed by somatic doubling early in development, giving rise to matroclinous and completely homozygous diploids. Diploid parthenogenesis, with the formation of a single polar body, and no reduction in chromosome number, is a normal occurrence in the life-cycle of aphids and many other animals.

PHENOTYPIC CHARACTERISTICS OF POLYPLOIDS

Haploids and autopolyploids are phenotypically different from diploids of the same species. In general, triploid and tetraploid plants are larger, darker green, and generally more vigorous than diploids. Triploid and tetraploid females of Drosophila are larger and have a more "robust" appearance than diploids. Haploids are characteristically smaller and less vigorous than diploids. Within an autopolyploid series, there is usually a more or less close relation between cell size and chromosome number. Thus, Navashin's measurements of dermatogen cells in Crepis capillaris show 1.8, 4.0, 6.0, and 9.0 respectively as the volumes (in 1,000 cu. μ) for 1n, 2n, 3n, and 4n plants. This difference in cell size is often taken advantage of in detecting polyploids phenotypically. Triploid females in Drosophila can be recognized by the size of wing cells, for example. The method has even been used in certain groups of plants by Sax and Sax for herbarium specimens by determining the density of stomata, which is related to cell size.

ALLOPOLYPLOIDY

An interspecific hybrid in which one parent was an autotetraploid and the other a diploid has two complete sets of chromosomes from one species and one set from the other. The chromosome sets from different species that are present in such an allotriploid need not necessarily be alike in the number of chromosomes present. As an example of such an allotriploid, the hybrid between Zea mays (maize; $n = 10$) and an autotetraploid form of the related species Tripsacum dactyloides ($n = 18$) may be cited. This intergeneric hybrid, which can only be obtained by the use of special methods, has 46 chromosomes—two sets of 18 from Tripsacum and one set of 10 from maize. In this case, studied by Mangelsdorf and Reeves, as in many similar ones, the homologs in the two like sets from Tripsacum pair to form 18 bivalents and the 10 maize chromosomes remain as univalents. The univalents are apparently distributed at random at meiosis, whereas the bivalents are distributed regularly, one set to each pole.

Raphanobrassica.—In a great many hybrids between diploid species of plants the chromosomes are sufficiently dissimilar that there is little or no pairing of chromosomes at meiosis and only univalents are formed. As a consequence of the irregular distribution of these univalents, such hybrids are highly sterile. A classical example of a hybrid of this type is that between the radish

(Raphanus sativus) and the cabbage (Brassica oleracea). Each of these species has nine pairs of chromosomes. The hybrid, known as "Raphanobrassica," has been studied by Karpechenko. It is remarkable for its vigor (see discussion of heterosis, page 276), and is only slightly fertile. At meiosis there is little or no chromosome pairing. An occasional seed is set, however, and

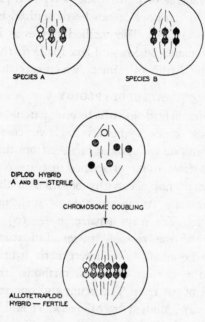

Fig. 107.—The production of an allotetraploid. Species A and species B, each with three pairs of chromosomes, produce a sterile diploid hybrid. Doubling of the chromosomes then produces a fertile allotetraploid, with six pairs of chromosomes, three from A and three from B.

most of these arise from the fusion of unreduced eggs with unreduced sperms. Plants from such seeds are fully fertile. At meiosis there are eighteen pairs of chromosomes. Each chromosome of radish and each of cabbage is present in duplicate, and normal pairing occurs—radish chromosomes pair with radish chromosomes and cabbage chromosomes with cabbage chromosomes (Fig. 107). Thus, although the plant is a tetraploid, it possesses only pairs of chromosomes which are segregated at meiosis in the same manner as is characteristic of a diploid. Such an allotetraploid hybrid is not only fully fertile but is true-breeding. The characters by which the radish and cabbage differ have no opportunity to segregate because radish and cabbage chromosomes never pair.

Nicotiana tabacum.—A large number of true-breeding allotetraploid plant hybrids like that between the radish and the cabbage have been produced experimentally. They are often called *amphidiploids* or *double diploids*. One example, studied especially by R. E. Clausen and collaborators, represents a new origin of the common cultivated tobacco (Nicotiana tabacum). This species has 24 pairs of chromosomes while many of its wild relatives have only 12 pairs. It was, therefore, early thought to be an allotetraploid. In hybrids with the 12-chromosome form, N. sylvestris, there are 12 bivalents and 12 univalents, suggesting that one of the sets present in tabacum came originally from this species or one very much like it. Likewise in hybrids of N. tabacum and N. tomentosiformis there are 12 bivalents and 12 univalents, which suggests that the second set in tabacum came originally from tomentosiformis or from one of its very close relatives, of which several are known. Furthermore, the diploid hybrid between sylvestris and tomentosiformis shows low chromosome pairing and resembles tabacum in many important respects. Many attempts to obtain doubling in this sterile hybrid have been unsuccessful, but recently, by using the indole-3-acetic acid treatment, previously mentioned, allotetraploids have been obtained. While studies of these allotetraploid hybrids are still in progress, it seems reasonably certain that Nicotiana tabacum, the cultivated tobacco, originated through the accidental doubling of the chromosomes in an otherwise sterile hybrid between two wild species—presumably a hybrid between species very similar to the now-existing species, N. sylvestris and N. tomentosiformis.

Segregation of Genes in Amphidiploids.—Any locus present in each of the parental species will be represented four times in an amphidiploid. Starting with normal alleles at these two loci, a recessive mutation at one of them would not be expected to show phenotypic effects, because of the presence of the normal allele at the second locus. A second mutation involving the second locus, however, would enable the character to show. Thus, differentiating the alleles at the two loci to indicate their independence in segregation, we can have any one of the following conditions:

	Locus 1	Locus 2	
(1)	$+/+$	$+/+$	Normal—no segregation
(2)	$+/+$	$+/a_2$	Normal—segregation for gene a_2 but without phenotypic effect
(3)	$+/+$	a_2/a_2	Normal—no segregation
(4)	$+/a_1$	$+/+$	Normal—segregation for gene a_1 but without phenotypic effect
(5)	$+/a_1$	$+/a_2$	Normal—segregation for genes a_1 and a_2 giving 15:1 phenotypic ratio
(6)	$+/a_1$	a_2/a_2	Normal—segregating for a_1 giving a 3:1 phenotypic ratio
(7)	a_1/a_1	$+/+$	Normal—no segregation
(8)	a_1/a_1	$+/a_2$	Normal—segregating for a_2 giving a 3:1 phenotypic ratio
(9)	a_1/a_1	a_2/a_2	Non-segregating—mutant phenotype

It is evident, then, that in order to have a recessive mutant character show there must occur two independent mutations of the gene involved. Furthermore, 15:1 ratios might be expected to occur rather frequently in allopolyploid forms. This is apparently the case; many 15:1 ratios are found in species of this type. This does not mean, of course, that 3:1 ratios for recessive mutant characters to not occur in such forms. Inspection of the possibilities enumerated above shows at once that if a recessive gene mutation occurs at one locus and becomes homozygous before a similar mutation occurs at the second locus, the second mutation will, on segregation, result in a 3:1 ratio. While 15:1 and 63:1 ratios are not necessarily diagnostic of allopolyploidy, they do raise a presumption of this condition. Ratios of 15:1 do occur in species for which there is little evidence of polyploidy, and Sprague has further shown that two heterozygous gene pairs concerned with scutellum color in maize may give either a 9:7 or a 15:1 ratio, depending on whether a third gene is homozygous for its recessive allele or for its dominant allele.

Monosomic Types.—As might be expected, spores carrying one less than the diploid number of chromosomes may give viable gametophytes in allotetraploid plants. Consequently, it is possible to maintain strains in which some plants are deficient in one entire chromosome. Such plants are, strictly speaking, 4n — 1 in constitution, but, because of the diploid-like character of allotetraploids they have been referred to as monosomic types. A number of such monosomic types are known in Nicotiana tabacum (Clausen), and they are useful in various ways in genetic studies. For one thing, they have distinct phenotypic effects on the plant. In N. tabacum there are 24 monosomic types possible. Of these, at least 20 are known. The monosomic type in tabacum known as "fluted," may be used as an example of the behavior of these types. Fluted tabacum crossed with tomentosiformis, one of the supposed parents of tabacum, gives (in the fluted hybrids) 11 bivalents and 13 univalents, whereas in crosses of fluted tabacum and the other supposed parent, sylvestris, fluted hybrids are obtained with 12 bivalents and 11 univalents. This shows that fluted is deficient in a chromosome (the "F" chromosome) originally derived from the tomentosiformis parent. There is a well-known recessive mutant form of tabacum known as "mammoth." Mammoth tabacum on the appropriate genetic background has many broad leaves, and for this reason is extensively used in commercial tobacco culture. In crosses of fluted tabacum with mammoth tabacum, all the fluted plants in the F_1 generation are also mammoth. This shows immediately that mammoth is normally carried in an "F" chromosome. The principle made use of here is the same as that of sex-linkage or of sectional deficiency.

Spartina townsendii.—There are many cases of the origin under controlled conditions of true-breeding amphidiploid hybrids similar to those discussed (see Darlington's Cytology for lists of these). Wild species of plants as well have originated in this way. One of the clear cases is that reported by Huskins for cord

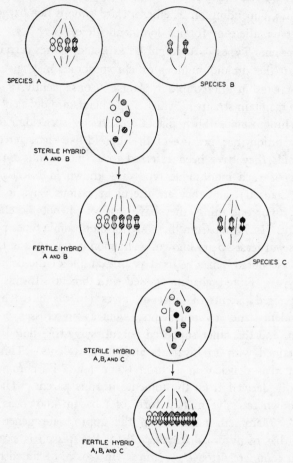

Fig. 108.—The production of a fertile allopolyploid combining the complements of three original species. The diploid hybrid between A and B undergoes doubling, and the resulting fertile hybrid crosses with C. In this way is produced a sterile hybrid with the haploid complement of each original species; doubling of its chromosomes again gives a fertile hybrid.

grass, Spartina townsendii. This species was not recorded by systematists until 1870, at which time it was known to occur in two small patches on the south coast of England. Taxonomically, it combines the characters of both S. alterniflora and S. stricta, and it was long believed by taxonomists to be a hybrid between these species. This supposition was supported by the fact that the two regions where it first appeared were just those in which the two suspected parents overlapped in their distributions. Huskins found the chromosome numbers of the sporophytes to be 70 in S. alterniflora, 56 in S. stricta, and 126 in S. townsendii. These chromosome numbers leave little doubt as to the correctness of the supposition that the species arose by hybridization followed by chromosome doubling. The new species has been very successful, has extended its range widely along the shores of the English channel, and has largely replaced its parental species wherever it has come into competition with them.

Spartina townsendii is probably not a simple allotetraploid. Both its parents, judging from chromosome numbers, are probably allopolyploids themselves. Most cultivated wheats are without much question also complex allopolyploids. Some wild species have seven pairs of chromosomes, others have fourteen, while the common cultivated wheats have twenty-one pairs. The latter are allohexaploids. The origin of true-breeding polyploids of this type is indicated in figure 108 as occurring in two steps, first, the formation of an allotetraploid hybrid between two species followed by the secondary occurrence of an allohexaploid hybrid between the allotetraploid and a third diploid species. The wild species ancestral to the modern cultivated wheats are not known with certainty, but a number of true-breeding allopolyploid hybrids involving wild species have been artificially produced.

Allopolyploidy and New Forms.—In the production of new forms of cultivated plants polyploidy has been of the greatest importance. In the hands of the scientifically trained plant breeder the techniques of inducing polyploidy will undoubtedly continue at an increasing rate to make possible the development of new varieties of crop plants, fruits, vegetables, and ornamentals. In sexually propagated species such as wheat and oats, hybrid sterility can be eliminated by making use of allopolyploids. In many fruits and ornamentals which are reproduced by asexual means the

sterility characteristic of autopolyploids and of polyploids intermediate between the two types is of no great disadvantage, while
the increased cell-size and plant-size often makes polyploids more
desirable than diploids.

Secondary Polyploidy.—An examination of the chromosome
numbers of groups of related species suggests at once that in
addition to change in number by the simple difference of entire
sets, chromosome numbers may change in other ways. Judging
from the organization of chromosomal material in related species,
it is evidently possible for two rod-shaped chromosomes to fuse
to give rise to a single V-shaped chromosome. The reverse of
this, namely, breakage of a V-shaped chromosome into two rod-
chromosomes each with a centromere is also a possibility. It is
entirely possible that changes in number other than those involving whole sets of chromosomes may be combined with polyploidy.
There is circumstantial evidence of this in some species. Such a
condition is often referred to as *secondary polyploidy*.

Polyploidy in Animals.—Habitual polyploidy is rare among
animals. This is readily accounted for by the fact that the usual
mechanism of sex-determination immediately breaks down in
polyploid individuals. This has already been made clear in
Drosophila. In such species as this in which the male is heterogametic, polyploid females are sexually normal but the sex mechanism in the male no longer operates successfully. The reverse
situation of course applies to forms in which the female is the
heterogametic sex. As a consequence, regular polyploidy occurs
only in animals that are hermaphroditic (no one satisfactorily
demonstrated instance, but several probable ones) or that reproduce parthenogenetically. Among the latter there are three well-
known examples: (1) Artemia salina, a brine shrimp, (2)
Trichoniscus provisorius, a sow bug, (3) Solenobia pineti and
S. triquetrella, bag-worm moths. In all of these species both
diploid and polyploid forms are known, and in each case the
diploid form reproduces sexually while the polyploid form reproduces parthenogenetically. Only females are produced in the
parthenogenetic forms. In the bag-worm, Solenobia, for example,

the sexual diploid forms have 30 pairs of chromosomes while the parthenogenetic form is a tetraploid. All of these parthenogenetic forms have circumvented the difficulties attending polyploidy by resorting to a form of asexual reproduction.

REFERENCES

Darlington, C. D. 1937. Recent advances in cytology. 671 pp. Blakiston's Sons Co., Philadelphia.

Dobzhansky, T. 1937. Genetics and the origin of species. 364 pp. Columbia University Press, New York.

PROBLEMS

1. A plant heterozygous for the two genes a and b gives a 9:3:3:1 ratio when self-fertilized. Its chromosomes are doubled to produce an autotetraploid of the constitution $+/+/a/a\ +/+/b/b$. Assuming no crossing over between these genes and the centromeres, and complete dominance, what ratio will such a plant give when progeny are produced by self-pollination?

2. What ratio will an autotetraploid plant give among its offspring produced by self-pollination if it was derived by simple somatic doubling of the chromosomes of a diploid that give a 9:7 ratio of selfing? Make assumptions similar to those indicated in the preceding problem.

3. If the locus of gene a gives no crossing over with the centromere, how could one get a plant of the constitution $a/a/a/a$ from an autotetraploid of the constitution $+/+/+/a$? Assume the organism to be a plant which can be either self-pollinated or outcrossed. For each step of the process give the proportion of plants expected of the constitution necessary for the succeeding step.

4. In a species of plant that can be either selfed or outcrossed, how could you distinguish an autotetraploid of the constitution $+/+/+/a$ from one of the constitution $+/+/+/+$? Give ratios wherever your proposed method makes use of these. Assume gene a to give no crossing over with the centromere.

5. Starting with three related diploid species of plants, outline the procedure you would use to get a true-breeding allohexaploid with chromosomes that formed simple pairs at meiosis.

6. East obtained a haploid tobacco plant (sporophyte), from which, through the production of viable haploid gametes, he obtained a completely homozygous diploid strain. This strain was at first

extremely uniform—each plant was exactly like each other one; but on continued sexual reproduction the uniformity gradually became less. Why?

7. Stadler measured the frequency with which mutant types were obtained in progeny obtained by self-pollination of x-ray treated diploid, allotetraploid and allohexaploid species of wheat. He found this frequency to be highest in the diploid and lowest in the allohexaploid species. How would you interpret these results?

8. In crosses between species of wheat with different chromosome numbers it is often found that reciprocal crosses give F_1 seeds that differ in development of the endosperm; the cross in which the female parent has the higher chromosome number is likely to be better than its reciprocal in this respect. Can you suggest any genetic basis for such a difference?

CHAPTER XX

SPECIES DIFFERENCES

GENETICS is concerned with the origin, development, and inheritance of differences between individuals. One of the most obvious and familiar classes of differences is that which distinguishes different species. One does not have to be a biologist to perceive the existence of species; language itself recognizes them. One has only to think of the associations that go with the words cat, lion, tiger, leopard, puma, jaguar, to form a picture of specific diversity—for each of these words is the name of a species in the single genus *Felis*.

Systematics.—Systematic biology is concerned with the recognition, naming, and describing of species, together with their habits and geographical distribution. Like some other branches of study, it suffers from the awkward circumstance that no two authorities can agree on a definition of the fundamental unit, species, with which it deals. Most biologists who have studied the matter are convinced that the species concept is a valid and useful one; but they cannot agree on a definition—and often can come to no working agreement with respect to particular examples. Some groups of organisms—such as the sedge genus Carex, the American forms of Aster, or the parasitic flies of the family Tachinidae—are notoriously difficult for the taxonomist; others are relatively simple—though in the most orthodox of groups there is always likely to be a section where species are not easy to delimit.

Discussions of this question have filled many volumes, and more are being written every year. Most of the recent ones are based in part on the results and principles of genetics. We shall describe some of the contributions that genetics has to make, especially with respect to two subjects: the ways in which related species differ and agree in their genetic constitutions, and the nature and inheritance of interspecific sterility.

SPECIES HYBRIDS

When two distinct species are crossed, the hybrids are usually partially or completely sterile. We shall take up first the genetic results obtained in cases where the sterility of the F_1 is not complete. In a few cases, certain specific differences have been shown to depend on differences in a single locus. East has shown that Nicotiana langsdorfii and N. alata differ by many genes. Most of the differences are of multiple gene nature and are therefore difficult to analyze in detail. Among the characteristic differences, however, there are two, each of which is differentiated by a single gene pair. One of these controls flower color and the other is concerned with pollen color. In a similar way, Detlefsen found that the guinea pig, Cavia porcellus, differs from the wild C. rufescens in many gene pairs, one of which determines the color of the coat on the ventral side of the body. In these and similar cases, it should be emphasized that these specified genes are not the only ones in which the species differ—they are merely the ones that give simple results. Actually such differences are exceptional; the rule is that specific differences are dependent on many genes, each of which has a slight phenotypic effect. Most fertile species hybrids give in F_2 extreme examples of the kind of multiple gene segregation described in Chapter XVII.

Maize-teosinte Hybrids.—Another example showing the kind of inheritance typical of characters in which species differ is found in hybrids between maize (Zea mays) and annual teosinte (Euchlaena mexicana) studied by Collins and Kempton, by Mangelsdorf, and by others. Each of these plants has 10 pairs of chromosomes and the hybrid between them is not highly sterile. The two species are very different morphologically. Maize has the complex pistillate inflorescence familiar to all as the "ear." Teosinte has a pistillate inflorescence made up of two rows of seeds imbedded in an alternate manner in bony joints of a simple rachis (Fig. 109). There are other differences, and the F_1 hybrid is more or less intermediate for most of them. One of the differences, that of whether pistillate spikelets are borne singly or in pairs, may possibly be due to a single gene pair. The F₂ generation of this hybrid gives a variety of intermediate forms, a small percentage somewhat maize-like and others more like the teosinte parent. This hybrid is unusual in one respect—the two parents are put in separate genera—and it is not often found that intergeneric hybrids are as fertile as is this one. This fertility of

Fig. 109.—"Ear" of teosinte (Euchlaena), with a single kernel of maize for comparison.

the hybrid may be taken as an indication that the two forms are not as widely different as might be supposed from the morphological differences. They perhaps should not be put in separate genera.

Drosophila melanogaster-simulans Hybrids.—In cases where the F₁ hybrid is completely sterile it is sometimes possible to compare the genetic constitution of two species. One method is illustrated by the comparison between Drosophila melanogaster and D. simulans. A series of recessive mutant types has been found in each of these species. If a recessive, of either species, is crossed to the wild-type of the other species the resulting F_1 is wild-type. That is, each species carries the dominant alleles of the recessive mutant genes known in the other. Many of the types found in the two species are phenotypically similar; allelism of the genes concerned may be tested by crossing. In melanogaster there are four similar recessive eye-colors—vermilion, cinnabar, scarlet, and cardinal. Two types of this kind are known in simulans. Crossing the homozygous mutant strains of simulans to each of the four in melanogaster gives these results: one type gives the characteristic bright-eyed mutant phenotype in F_1 when crossed to vermilion, wild-type when crossed to cinnabar, scarlet, or cardinal; the other gives the mutant phenotype when crossed to scarlet, wild-type when crossed to any of the other three. Evidently, then, the two mutant genes in simulans are allelic to vermilion and to scarlet, respectively. Tests of this kind have established the allelism of mutant genes in 29 different loci in each species. The linkage maps show these 29 loci to be distributed in much the same way in the two species; there are the same four linkage groups, with no exchanges of materials. The sequences within each group differ only in that simulans has a long inversion in the right limb of chromosome 3, as compared to melanogaster; there are only minor differences in the frequencies of crossing over in the various regions of the three long chromosome pairs. These genetic results have been confirmed by an examination of the salivary gland chromosomes of the F_1 hybrids. The large inversion in chromosome 3 is easily observed; there are also a few other short sections that differ, in ways that are not yet clear. Otherwise the salivary gland chromosomes do not suggest the hybrid nature of the flies from which they came.

These two species are very similar morphologically as well as genetically and cytologically. Their mitotic chromosomes show no differences, and it takes practice to distinguish the adult flies. The detailed resemblance in their genetic makeup is apparently exceptional, if one may judge from other, less complete comparisons that are possible in Drosophila.

Drosophila pseudoobscura-miranda Hybrids.—D. pseudoobscura and D. miranda give an almost completely sterile hybrid, the salivary gland chromosomes of which have been studied by Dobzhansky and Tan. The cytological picture is very complex here, and quite different from that encountered in any intraspecific hybrid. It appears that 50 or more breakage-points with subsequent rearrangements must be supposed to have occurred since these species diverged. There are about five short pieces deleted from one chromosome arm and transferred to another; the remaining rearrangements are inversions, each within a single arm.

More Remote Species.—These two species are still closely related; when one studies more distantly related forms of Drosophila, both the indicated methods of comparison become useless. Such remote forms cannot be crossed, so allelism of mutant genes and cytology of the hybrid are both impossible to study directly. Comparison of the structure of the salivary gland chromosomes is also unavailing; D. melanogaster and D. pseudoobscura are, in this respect, the two best-known species—and not a single section of the chromosomes of one of them has yet been identified in the other.

Chromosomes of Drosophila Species.—The chromosome configurations found in a series of species of Drosophila are shown in figure 110. Inspection of this figure indicates that rather extensive changes have occurred in the evolution of these species. Much of the diversity appears to concern the heterochromatin; melanogaster, pseudoobscura, virilis, funebris, and hydei may be supposed to differ only in heterochromatic material and centromere attachments; this is certainly too simplified an interpretation, but there are grounds for concluding that it represents a first approximation to the facts.

Parallel Mutations.—Mutant types have been studied in most of the species illustrated, and many of these are closely similar. As shown above, such similarities between the types of melanogaster and simulans can sometimes be shown to be due to allelism. When the species compared cannot be crossed, such comparisons become much more dubious; nevertheless, one striking result is that mutant types, in every species of Drosophila studied, that show clear and unambiguous resemblances to known sex-linked types of melanogaster, are themselves sex-linked. There is a single clear exception to this rule—bobbed in ananassae is auto-

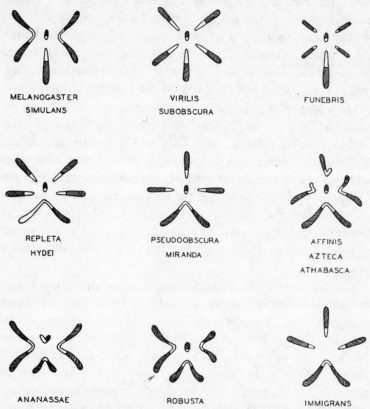

Fig. 110.—Haploid chromosome groups of a series of species of Drosophila. In every case the lower chromosome is the X. Shaded regions are "active" or euchromatic, unshaded are heterochromatic. The proportions of the several arms and the extent of the heterochromatin are only approximately correct.

somal. It may be concluded that the genes that lie in the X of melanogaster also lie in the X's of the other species. The converse of this rule does not hold; comparison of mutant types agrees with the cytological picture in suggesting that the X of pseudoobscura includes a whole arm that is autosomal—probably the left limb of the third chromosome—in melanogaster. It must be emphasized that the comparisons with which we are here concerned are rather speculative, and are not easily subject to experimental verification. Their justification is merely that they have not yet led to serious inconsistencies.

Differences among Drosophila Species.—It can be concluded that inversions are usual, translocations occasional, and fusions and breakages at or near centromeres also occasional, in the course of events leading to the differentiation of species of Drosophila. Gene mutations must also be assumed, and to them are probably to be attributed the phenotypic differences on which specific descriptions are based. Their number cannot be estimated in Drosophila, owing to the sterility of the hybrids.

INTERSPECIFIC STERILITY

One of the implications of the term species is that groups so designated represent relatively isolated populations, the genes of which are not freely interchanged. There are many isolating mechanisms that serve to prevent such interchanges. A rough classification of these mechanisms follows:

1. The two groups of individuals do not cross, for one of the following reasons:

    (*a*) Geographical—*i.e.,* they occupy different areas and never have an opportunity to cross.

    (*b*) Mechanical—*i.e.,* the sizes of the animals or the shapes of the parts make mating impossible.

    (*c*) Psychological—*i.e.,* sexual stimulation is not sufficient to induce cross-mating.

    (*d*) Physiological—*i.e.,* such relations occur as in the case of plants in which the pollen tube of one species will not grow in the style of another.

2. The hybrid zygote is formed, but is inviable. This is the relation found in such crosses as that between the rat and mouse, or fowl and turkey; in both cases embryos are formed, but die at an advanced stage. In other cases death occurs quite soon after fertilization.

3. Mature hybrids—often especially vigorous—are produced, but are sterile. The mule is the most familiar example of this type. The sterility of the F_1 hybrid may itself be due to a variety of causes; two main types may be distinguished:

(*a*) Failure of regular pairing and segregation at meiosis, there being no system of definite pairs of chromosomes present. This case has already been discussed in Chapter XIX, where it was shown that doubling the number of chromosomes remedies the difficulty and leads to the production of a fertile allotetraploid, in which pairing and meiosis are again regular.

(*b*) What has been called genic sterility also occurs. In this case doubling of the chromosomes does not lead to fertility, since the difficulty does not lie in failure of conjugation. In some such cases (*e.g.,* many bird hybrids, and the melanogaster-simulans hybrids referred to above) the degeneration of the germ-cells sets in before meiosis begins.

Primula kewensis.—One of the earliest cases studied cytologically furnishes a curious modification of the two types just discussed. The two primrose species, Primula verticillata and P. floribunda, each have nine pairs of chromosomes; their hybrid, known as P. kewensis, was obtained as a sterile diploid, propagated vegetatively. Eventually a tetraploid branch appeared, and this was at once seed-fertile. The tetraploid strain is regularly grown as an ornamental. The unusual circumstance in this case is that pairing and meiosis appear quite normal in the diploid hybrid —the sterility of which is therefore unexpected. It remains uncertain whether the cause lies in a developmental difference between the diploid and tetraploid plants—*i.e.,* the case belongs in subdivision "b" above—or whether independent segregation of the nine normally paired chromosomes in the diploid gives so many inviable gametophytes that their degeneration spoils the few good ones that must be supposed to be present.

Genic Sterility.—"Genic" sterility is evidently due to the action of complementary genes, derived from the two parents. Only in this way can one picture the formation of a sterile F_1 from the crossing of two strains each of which breeds true for fertility. There exist cases in which one of the parent strains is not homogeneous for such a complementary gene; in these cases the analysis can be carried out.

One example of this general nature was studied by Hollingshead in Crepis. If C. capillaris is crossed to certain strains of C. tectorum, the hybrids are fully viable; if other strains of tectorum are used the hybrids are extremely abnormal and do not develop beyond the cotyledon stage. Crosses between the two strains of tectorum used give no aberrant plants, either in F_1 or F_2; but tests show that the difference in reaction of the two kinds of hybrid plants is due to a single pair of alleles, for which tectorum is heterogeneous. These alleles are without any detected phenotypic effect in tectorum.

Races of Drosophila pseudoobscura.—When interspecific sterility is complete, it is, in the nature of the case, impossible to study its inheritance. The inheritance of incomplete sterility has, however, been investigated. Perhaps the most thoroughly analyzed example is that of the two "races" of Drosophila pseudoobscura, discovered by Lancefield and more recently studied by Dobzhansky and others. This species, common in western North America, exists in two races, called A and B, that have some of the properties of distinct species. Individual specimens cannot be distinguished by eye, but there are slight average differences in structure and more definite ones in physiological properties—especially in that race A is relatively more tolerant of high temperature. The geographical distribution is different, B being limited to the northwestern part of the area, A occurring in most of this region and in a much wider one as well. The striking difference, however, is this: if the two races are crossed, all the F_1 males are sterile. There are two proofs that this sterility is genic, and not due to absence of chromosomes sufficiently alike to pair: occasional tetraploid cysts occur in the hybrid testes, and do not form normal sperm; and the F_1 sisters of these same males are reasonably fertile. This latter point has made it possible to carry the analysis further, by studying the offspring of the A-B hybrid females when back-crossed either to A or to B males. The genetic basis of the sterility is found to be complex. There are at least three major differences and several minor ones, of such a nature that fully fertile males are produced only when all alleles in the loci concerned were derived from one of the two parent races. Cytologically, the two races differ in several inversions; but only three of these are invariably present. Greater differences than this occur within race A without leading to any detectable sterility. Here, then, we have direct evidence that sterility is dependent on complementary genes.

The origin of specific differences has been one of the most discussed topics in biology for eighty years—since the publication of Darwin's "Origin of Species." We have outlined the evidence as to the nature of the differences that exist; both gene differences and differences in chromosome organization (inversions, translocations, polyploidy, etc.) occur. Given an initial separation into populations that do not interbreed, the accumulation of such differences is not difficult to imagine. The real difficulty is in picturing the origin of the isolating mechanism that produces two such separate populations.

One solution of this problem has already been given—that of allopolyploidy. This process leads to the production at once of a new and fertile type. Such a type is commonly isolated from the very beginning; tetraploid Raphanobrassica, for example, crosses only rarely either with radish or with cabbage. Even if seeds are produced by an allotetraploid crossed with one of its diploid parents, the hybrids are usually triploid and highly sterile. There can be no doubt that this method of specific differentiation has played a large rôle in the evolution of hermaphroditic plants.

Polyploidy is not a likely mechanism of specific differentiation in dioecious animals, and cannot be supposed to be the only one at work in plants. It is difficult to picture in detail, but there can be no doubt that the complementary gene mechanism, now in process of development in Crepis tectorum and in Drosophila pseudoobscura, often lies at the basis of species differentiation.

REFERENCES

Anderson, E. 1937. Cytology in its relation to taxonomy. Botan. Review, 3:335–350.

Dobzhansky, T. 1937. Genetics and the origin of species. 364 pp. Columbia University Press, New York.

Fisher, R. A. 1930. The genetical theory of natural selection. 272 pp. Clarendon Press, Oxford.

Haldane, J. B. S. 1932. The causes of evolution. 234 pp. Harper and Brothers, New York.

Hertwig, P. 1936. Artbastarde bei Tieren. Handbuch Vererbungswiss., 2, B. 140 pp.

Sax, K. 1935. The cytological analysis of species hybrids. Botan. Review, 1:100–117.

PROBLEMS

1. Hairless in Drosophila melanogaster is a dominant mutant type, the gene having also a recessive lethal effect. A similar character, with the same properties, was found in D. simulans. How would you test the possible allelism of these two?

2. The relation between the attachments of the chromosome arms in D. melanogaster and D. pseudoobscura indicated in the text is based in part on the following parallel types in D. pseudo-obscura: yellow, miniature, bobbed, cinnabar, delta, white, scarlet, brown, star, echinus, vermilion, forked. How are these distributed in the linkage groups of D. pseudoobscura?

3. What is the difficulty in supposing that interspecific sterility arose by a single mutation, the two types AA and aa giving a sterile heterozygote?

CHAPTER XXI

EXTRACHROMOSOMAL INHERITANCE AND MATERNAL INFLUENCES

THE question may be asked: Is all heredity due to genes carried in chromosomes, or are other bodies also to be taken into account? In what follows we shall confine our attention to higher organisms, in which definite chromosomes are present; the situation in such organisms as bacteria or blue-green algae, in which chromosomes seem to be absent, is too confused to be discussed profitably—especially since these forms have no sexual reproduction, without which most of the geneticists' standard techniques cannot be used.

PLASTID INHERITANCE

In the higher plants it is clear that one set of bodies other than the chromosomes does carry its own independent series of inherited properties—namely the chloroplasts. Histological study shows that in seed-plants the chloroplasts develop from small bodies known as proplastids; there is some evidence that proplastids divide, but their origin is not altogether clear, owing to their small size and to the difficulty of distinguishing them from other granules. The genetic evidence, however, is less ambiguous.

Genes Influencing Plastids.—Variations in plastid color are common in seed-plants; we have already referred to white seedlings in maize, and various degrees of yellowness or paleness of the green color are also frequent in most seed-plants. The majority of these types are due to recessive gene mutations, and are inherited like other characters due to genes in the chromosomes. It is clear, then, that the color of the plastids is influenced by the genes. There are, however, other plastid variations, phenotypically similar to these, that are inherited only through the female line.

Maternally Inherited Plastid Characters.—Correns studied such a form in the Four-o'clock, Mirabilis jalapa. There is a strain, known as albomaculata, in which the leaves are irregularly mottled with green and yellowish white (Fig. 111). The difference in color is due to the color of the individual plastids, which are either dark green or very pale. Near the boundaries between two areas may be found cells in which both kinds of plastids are present. Evidently these are disposed at random at cell division, and the green and white areas represent descendants of cells which have, by chance, come to contain only one type of

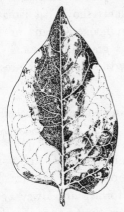

Fig. 111.—Variegated leaf of Four-o'clock (Mirabilis). The dark areas have normal green chloroplasts, the white areas have colorless ones. (After Correns.)

plastid. Such uniform areas may include whole branches. If the flowers on such branches are pollinated, the resulting seeds on the wholly green branches give rise only to green seedlings, those on the white branches only to white ones, which die because of lack of photosynthesis after they have used up the reserves present in the seed. Seeds from mosaic branches give a mixture of green, white, and mosaic seedlings. These relations hold whether the pollen be obtained from the mosaic plant itself or from a normal green plant. If a normal green is fertilized by pollen from any flower on the mosaic plant, only green seedlings are produced in F_1 and F_2. In short, the egg determines the nature of the plant produced, and the nature of the egg depends on whether it happens to carry green plastids, pale ones, or both. Cases quite like this are known in many species of plants.

Plastids in Oenothera Hybrids.—Renner has studied the genetic behavior of the plastids in Oenothera. It appears that each species has its own characteristic plastids, which retain their specific properties even when kept for many generations in plants that have genes characteristic of other species. Oe. hookeri is a California species, homozygous and true breeding; we have already referred to Oe. lamarckiana as a balanced heterozygote, made up of the two complexes velans and gaudens. Reciprocal crosses of these two plants show that lamarckiana plastids give pale green seedlings in hookeri-velans or homozygous hookeri plants. These seedlings gradually become darker green as they develop, and produce fertile plants. The results in F_1 are as follows:

hookeri $\times$ lamarckiana $\rightarrow$ green gaudens-hookeri, green velans-hookeri
lamarckiana $\times$ hookeri $\rightarrow$ green gaudens-hookeri, pale velans-hookeri

If the gaudens-hookeri F_1 plants are pollinated by hookeri, there result two types of plants—gaudens-hookeri and homozygous hookeri. The former are always green, the latter are green if the F_1 came from the first mating above, pale if it came from the second one—*i.e.*, if it carried lamarckiana plastids. Other matings confirm the conclusion stated above: Lamarckiana plastids give pale plants if the genes are of the kinds present in velans and hookeri, green plants in the presence of gaudens genes.

Oenothera differs from some other plants in that occasional plastids are brought in by the pollen. Random sorting of these during the early development of the embryo leads to the production of patches of tissue with paternal plastids. The result is that the F_1 pale plants described above often have patches of dark green tissue. Such patches are never present if the pollen parent carried lamarckiana plastids. When the dark green areas are present, they may include whole branches; the flowers on these behave, on crossing, as do whole plants with hookeri plastids.

There can be no doubt that, in Oenothera and in seed-plants in general, the development of the plastids is dependent on the interaction of their own inherent properties with those of the genes. The plastids are capable of indefinite reproduction; while their phenotype is partly dependent on the genes that occur with them, they have permanent properties that are maintained without regard to the genes.

Male-sterility in Maize.—There are other characters in plants that are inherited in this same way. Rhoades has described a "male-sterile" type in maize. The abnormality is transmitted solely through the female line; though some fertile pollen is produced, it does not transmit the male-sterility. Cytological study shows that the sterility is due to degeneration of the fully formed microspores, which differ from those of normal plants in the size, shape, and number of cytoplasmic elements that are probably plastid-like in nature.

Other Maternal Characters.—In some similar examples the character concerned does not have any obvious relationship to the plastids; but it remains possible that the maternally inherited element concerned is in the plastids. In order to establish the existence of another autonomous element involved in heredity, it is necessary to find cases in which the facts cannot be explained in terms of the two known elements—chromosomal genes and plastids. The exclusion of a possible plastid effect in green plants is difficult. Animals do not have plastids, and one may ask: Do they furnish evidence for the existence of any permanent autonomous elements outside the chromosomes?

MATERNAL EFFECTS

Rôle of the Cytoplasm.—Many embryologists have concluded that the cytoplasm is of primary importance in determining the nature of the individual, the chromosomes being concerned only in the differentiation of minor, relatively superficial characters. This view, advanced by Boveri, Conklin, and Loeb, among others, was formulated by Loeb in the statement that the cytoplasm determines the "embryo in the rough," the genes in the chromosomes being responsible only for minor modifications of a general ground plan determined independently of them. This view is based on the study of hybrids between widely different kinds of animals. It is possible to fertilize or activate eggs of sea-urchins by the sperm of crinoids, or even of molluscs. The resulting embryos do not develop far; but so far as they do go, they are sea-urchin embryos, without any trace of the characteristics of the species from which the sperm came. It may be doubted if the foreign sperm does anything more than induce parthenogenetic development in these cases; but in less wide crosses, where later development occurs and shows definite sperm influence, it is still true that the earliest stages are of the type characteristic of the species from which the egg came.

These and other experiments demonstrate that the cytoplasm of the egg contains the potentialities of early development; but they do not demonstrate that this property is due to self-perpetuating elements in the cytoplasm. It is probable that the genes in the chromosomes of the mother are ultimately responsible for all the specific properties of the cytoplasm, as can be shown in certain instances.

Coiling in Limnaea.—The fresh-water snail Limnaea normally has a dextrally coiled shell (Fig. 112), but races occur with sinistral coiling, *i.e.*, the shell is a mirror image of the usual type. The results of Boycott, Diver, and others, show that both types occur in true-breeding strains. If these are crossed, all of the F_1 snails show the same type of coiling as the mother—*i.e.*, dextral $\times$ sinistral $\rightarrow$ dextral, sinistral $\times$ dextral $\rightarrow$ sinistral. The animals are hermaphroditic, and isolated individuals reproduce by self-fertilization. F_2 families obtained in this way are made up entirely of dextral individuals, whichever type of F_1 is used. This result may be interpreted by supposing that there is a recessive mutant gene, s, for sinistral coiling. The direction of coiling of an individual is determined, however, not by its own genetic constitution, but by that of its mother—*i.e.*, by that of the diploid oöcyte from which the individual was derived. Thus a pure dextral strain is $+/+$, a sinistral one is s/s. All the F_1 individuals are $+/s;$ their coiling depends on which race the mother belonged to, but their offspring are all dextral. The test of the hypothesis comes from the behavior of the F_2 individuals when they are allowed to produce an F_3 generation by self-fertilization; three-fourths of them give dextral offspring, one-fourth give

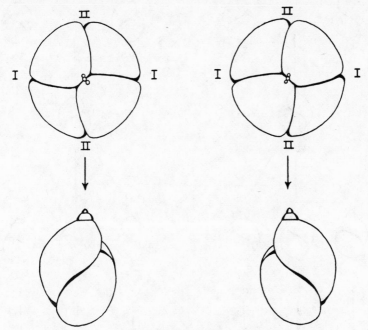

Fig. 112.—Cleavage and adult shells of snails. At left, dextral type; at right; sinistral. The three small bodies in the upper figures are the polar bodies; their position marks the upper pole of the egg. Given this and the distinction between the planes of the first (I) and second (II) cleavage furrows, the two types of eggs may be seen to be mirror images of each other, with respect to the direction of the cross furrow. This mirror-image relation persists throughout development, and is evident in the direction of coiling of the adult shells.

sinistrals (Fig. 113). It is thus possible for a phenotypically sinistral individual to produce either all sinistral or all dextral offspring, and the same two possibilities exist for a phenotypically dextral individual.[1]

Cases of this type are sometimes said to show *maternal inher-*

[1] There are certain families in which further complications occur; the descriptions given here and the interpretation apply to the majority of strains.

itance—a misleading term, since inheritance is perfectly normal, the unusual feature being that the phenotypic expression is delayed a generation. In the present instance such a relation is not surprising, for embryological study shows (Crampton, Kofoid)

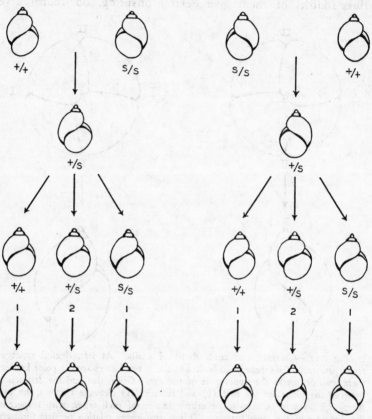

Fig. 113.—Results of reciprocal crosses between dextral (+) and sinistral (s) races of Limnaea peregra. The inheritance is normal, sinistral being recessive; but the direction of coiling of the shell of an individual is due to the constitution of its mother. The F₂ and F₃ generations are produced by self-fertilization.

that the two types of coiling are determined very early; they may be identified at the second cleavage. The relations here are illustrated in figure 112. The top of the egg may be identified by the presence of the polar bodies; given this and the distinction between first and second cleavage furrows, the direction of the "cross-furrow" is diagnostic.

The gene responsible for the direction of coiling may have been eliminated in a polar body (in the case of an F_2 dextral that will give F_3 sinistral offspring), but the time interval between the meiotic and second cleavage divisions is short—too short, as the results show, for the genetic composition established at fertilization to make over the pattern of cleavage.

Delayed Expression of Characters.—Other examples are known in which there is a delay in the expression of a phenotype, the result being that individuals may show a phenotype dependent on genes no longer present. One relation of this sort is rather frequently found in pollen grains. We have already described the case of the waxy type of maize, where the kind of starch developed in a pollen grain is dependent on the genes carried in the grain itself. There are, however, other pollen characters that behave differently. In sweet peas, for example, there are strains with long pollen grains, others with round ones. The F_1 plants produce only long grains, in F_2 there are three plants with long grains to one with round ones. Here, then, the character of the gametophyte is determined by the genes present in the sporophyte that produced it—*i.e.,* the shape of the pollen grains is already fixed before the completion of the meiotic divisions that lead to the formation of haploid microspores, and a heterozygote produces half its pollen grains showing the dominant phenotype but lacking the dominant allele.

CYTOPLASMIC INHERITANCE

In all these cases the characters described are clearly under gene control; the only complication is that there is a delay in the expression of the phenotype. These relations have no bearing on the original question from which their discussion arose—are there permanent autonomous elements other than chromosomes and plastids? A number of examples have been described in the literature which, taken at face value, do establish the existence of such elements. In none of these does the evidence seem to us conclusive. Some of them are based on too few individuals to be critical, others involve complex experiments in which there are many opportunities for unsuspected errors to creep in, and in still others the possibility that the observed effects may be due to plastids or to ordinary genes has not been excluded.

Failure to find evidence of such extra-nuclear, extra-plastid elements cannot be accepted as final proof that they do not exist. There are, however, so many experiments that have given negative results, that it seems legitimate to conclude that the existence of such bodies can be accepted only when clear and unequivocal proof is produced.

Acetabularia.—One of the most striking results in this field is that which Hämmerling has obtained with the unicellular green alga Acetabularia, found in the Mediterranean Sea. This form has its nucleus in a basal root-like portion; from this portion arises a stalk, which is surmounted by a mushroom-like cap. There are two species, that differ markedly in the structure of the cap. If growing stalks are cut in two, and reciprocal grafts are made between the two species, each grafted tip of stalk will form a cap—and this cap is found to have the characteristic form of the basal portion of the composite—*i.e.,* of the nucleus and not of the grafted stalk (with its contained chloroplasts) that actually forms the cap. Here, then, the nuclear control is practically immediate and complete.

Relation of Genes and Cytoplasm.—One may not conclude that the cytoplasm is of no importance in development. It is obvious that, if it were possible to dissect out a complete set of genes and place them in a vacuum, they would not produce an organism. The cytoplasm is not only a necessary medium for the growth of genes and for the production of their specific effects; it is itself specific. There is abundant evidence that the cytoplasm of different species, and even of different individuals, has a highly specific composition. The question at issue, however, is: Is the specificity of the cytoplasm to be referred back to the specificity of the genes, or is it a permanent property that reproduces itself regardless of the genes present? That is, do the genes modify the nature of the cytoplasm, and, given time, mould it into a specificity determined by their own properties? At present the most probable answer to these questions is: Only the chloroplasts are known to have permanent properties independent of the genes in the chromosomes.

REFERENCES

Correns, C. 1937. Nicht mendelnde Vererbung. Handbuch Vererbungswiss., 2, H. 159 pp.

East, E. M. 1934. The nucleus-plasma problem. Amer. Nat., 68: 289–439.

Haldane, J. B. S. 1932. The time of action of genes. Amer. Nat., 66:5–24.

Hämmerling, J. 1934. Über Genomwirkungen und Formbildungsfähigkeit bei Acetabularia. Arch. Entwick. Mech., 132:424–462.

Sirks, M. J. 1938. Plasmatic inheritance. Botan. Review, 4:113–131.

PROBLEMS

1. In breeding Limnaea it is sometimes difficult to determine whether an individual has mated with another one or has self-fertilized. There is a recessive gene for albinism in the species, which does not show the delayed expression characteristic of direction of coiling. Show, in detail, how this gene may be used to establish that cross-mating has occurred in experiments on the inheritance of direction of coiling.

2. Two different pale green types occur in a plant—one due to an inherent property of the plastids, the other dependent on a recessive gene. What is the result, in F_1 and F_2, of reciprocal crosses between the two pale types?

3. In certain marine animals it is possible to obtain fragments of eggs that have no nuclei. In some cases these fragments may be induced to develop without fertilization (*i.e.,* parthenogenetically). Such development without nuclei never proceeds far; but it has been argued that the occurrence of any development at all proves that the cytoplasm, rather than the nucleus, is the important element in heredity. Is this argument convincing?

CHAPTER XXII

GENES AND PHENOTYPES

THAT there must be a relation between genes and hereditary characters is so obvious that it is often taken for granted without serious thought. This is because the relations in most instances are more or less invariable and therefore *apparently* simple. That is to say, substitution of one allele of a gene for another in an organism, gives as the end result a simple change in a character. For example, the color of the coat in dogs is changed from black to brown following the substitution of one allele for another. We have considered many examples of "simple" characters of this kind in this book—thousands of them are known—and they naturally lead one to think of gene and character together, with little or no emphasis on what connects them in the processes of development. That the relations are not simple is at once evident when one pauses to consider them in detail. A mere enumeration of some of the characters that are influenced by genes is enough to make this clear. To choose a few examples more or less at random—genes are known that, under given conditions, control such diverse characters as dormancy of seeds in plants, chlorophyll development, flower colors, ability of blood to clot in man, disease susceptibility in both plants and animals, mental ability in man, and many others. In short, in a specified environment, genes determine what kind of an individual a representative of a given species is going to be. There can be little doubt that genes also determine to what species a given individual will belong. By logical extension, it can be argued that genes determine whether an organism is a plant or an animal, as well as what kind of a plant or animal. And, to carry these deductions still further, genes determine whether or not an organism is going to develop at all.

There is no question that genes determine the nature of developmental reactions and the direction in which they will lead—but what can be said about *how* genes do these things? This question of how genes act is one of the major unsolved problems in biology. It is, of course, inseparably tied up with the general problem of differentiation, a problem in which only recently have profitable leads been opened.

Gene-character System.—To begin with, developmental reactions—reactions with which genes must be assumed to be concerned—form a complex integrated system. This can be visualized as a kind of three-dimensional reticulum—that is, many different but related reactions are going on simultaneously. It can be assumed that reactions start with the immediate products of genes. These products then interact to give secondary products and so on. From such a picture as this it should follow that genes which influence reactions that come early in the network would influence many subsequent reactions and consequently, when they mutate, result in drastic modifications of the system, *i.e.,* have a good chance of being lethal. Mutant characters that are non-lethal, then, should, in general, be differentiated by genes that influence reactions that either come late in development or that form part of a sub-network or chain of reactions of less physiological importance; for example, ones having to do with certain pigment systems.

Nature of Genes.—Since one is forced to assume that either genes themselves or immediate products of gene activity take part in developmental reactions, it is pertinent to inquire into the nature of genes. Here, too, there is little in the way of facts to go on, and one is forced to resort to speculation. A reasonable supposition is that genes either are proteins or are associated with proteins. In size they are of the order of large protein molecules and it is therefore conceivable that they are single large molecules. On the other hand, they may be aggregates of smaller molecules. Since genes are small in size and appear to be permanent (*i.e.,* not used up during development), it has several times been suggested that they might act directly as enzymes in catalyzing reactions, or might produce enzymes as immediate products.

Method of Direct Study of Genes.—It is evident from purely *a priori* considerations that it is possible to attack the problem of the relation of gene and character from either the gene end or the character end of the reaction system. In practice, however, it is usually difficult to start at the gene end. The gene itself is an unknown quantity and the number of ways of finding out more about it have so far remained strictly limited. A study of induced mutation (Chapter XIII) provides one way of finding out something about genes, but this method has several limitations, the most evident and serious one being that changes in a gene are inferred only from changes in the character; and without a knowledge of the connecting reactions, this gives little information as to the nature of the gene itself. Another possibility of promise is that of direct examination of chromosomes. By means of ultra-violet absorption methods, Caspersson has shown, among other things, that nucleic acids are an important constituent of chromosomes (as was already known) and that these are concentrated in the dark bands of the salivary gland chromosomes of Drosophila. One of the difficulties here is that there is reason to suspect from size considerations that a band of a salivary chromosome contains a high proportion of extragenic material and a low proportion of actual genes. The phenomenon of position effect (Chapter XIV) offers the hope of finding out something about immediate gene products, since it is presumably the interaction of these that is responsible for the phenomenon. Here again, however, the only available index remains the final character.

Dosage Relations.—Still another method of obtaining information about gene action involves a study of variations in the number of representatives of a given gene that are present, *i.e.,* dosage relations. As an example of this type of approach, the measurements of vitamin A content of maize kernels made by Mangelsdorf and Fraps may be mentioned. Maize with white endosperm contains very little vitamin A because of the small quantity of its yellow carotenoid precursor. Yellow endosperm varieties contain both carotenoid pigment and vitamin. Since the endosperm of maize is triploid, it is possible to obtain kernels, the endosperms of which carry 0, 1, 2, or 3 dominant Y (yellow) alleles and correspondingly 3, 2, 1, or 0 recessive alleles. It was found that the relative number of rat units of vitamin A was 0.05, 2.25, 5.00, and 7.50, respectively, for the four types. Such a direct proportionality suggests that the relation between gene and vitamin may be relatively simple. It does not tell what the gene is, but indicates that in this particular instance it is probably not acting as an enzyme—concentration of an enzyme does not usually show a linear relation to the amount of product formed by a reaction that it catalyzes. A general difficulty with methods involving dosage relations is that it is not usually known how the recessive gene acts. In the example just cited, the simplest assumption is that the recessive gene is doing nothing, but, of course, this is not a necessary assumption. In this connection, comparisons of the behavior of a recessive allele of a given gene and a known deficiency for that gene, do often suggest that, with respect to a given reaction chain, the recessive allele is actually inactive.

With regard to dosage relations, a fourth chromosome recessive character in Drosophila, shaven (*sv*—bristles and hairs removed) is of interest. Haplo-4 shaven shows the character in extreme form. Diplo- and triplo-4 flies, each with only *sv* alleles, show the character in progressively less extreme form. Tetra-4, usually inviable, is according to Schultz, who has studied this series, viable when only *sv* is present, and is very nearly wild-type phenotypically. It appears that the recessive allele is, in this case, doing the same thing as a normal allele but less efficiently. Four recessive alleles are approximately equivalent to two normals. A general objection to this type of experiment is of course that in the series with one, two, three, and four fourth chromosomes other genes at loci other than that of *sv* are multiplied, *i.e.*, the number of *sv* alleles is not the only variable in the experiment. Disregarding this difficulty, it is clear that the evidence may lead to deductions as to *how much* of something a given allele of a particular gene is doing, but has no bearing on the more important question of *what* it is that is being done.

The above examples serve to illustrate a few of the methods by which we may eventually hope to learn more about the nature of genes and their actions by more or less direct approaches at the gene end of the gene-character chain. That more progress has not been made is perhaps discouraging, but we may anticipate that these techniques will in the future be used more efficiently and that new ones will be developed.

Studies of Characters.—Technically it is much easier to work from the final character back toward the gene than it is to do the reverse. This amounts to describing the character in more and more detail. A complete description of any character would, of course, be a description of all of development, including the rôles played by all the genes present. Although such an ideal description is out of the question, one might hope to build up a partial description of a character which would include at least one unbroken sequence of reactions back to a gene. From such a sequence one would be able to determine at least some of the properties of the one gene known to be involved.

Many methods are being brought to bear on the problems of detailed character description. All we can reasonably hope to do in the present discussion is to describe briefly a few of the more or less typical examples of what is being done.

Coat Colors in Mammals.—The inheritance of coat colors is well understood in a number of mammals, particularly in various species of rodents. Wright has worked extensively with coat-colors in guinea-pigs and has attempted to determine in some detail the type of interactions obtained. The pigments involved are melanins, which are known to be produced by reactions catalyzed by enzymes. These reactions involve tyrosin or related amino acid-like substances (chromogens), and tyrosinase or similar enzymes are concerned. The final colors are the result of several conditions: (1) the kind of pigment produced, *e.g.*, black or yellow, (2) the total quantity of pigment, (3) the distribution of the pigments within individual hairs of the animal, and (4) the distribution of variously pigmented hairs over the body of the animal. With these variables, a large number of final combinations are, of course, possible. Wright has postulated that three substances, immediately preceding the development of pigment, are concerned. One is presumably the chromogen (tyrosin or a similar substance), the remaining two may be assumed to be oxidizing enzymes like tyrosinase. Following additional specific assumptions, it is possible to interpret the observed facts by supposing that both the initial presence and the quantities of the two enzymes, or enzyme precursors, are under the control of genes. The albino series of multiple alleles, for example, are presumably concerned with the production of substance I (enzyme or enzyme precursor). The various members of this series differ in such a way that they determine different amounts of this substance varying from none to full quantity. Presumably the quantity of substance I produced bears a rather direct relation to the nature of the alleles of the a gene. This admittedly speculative interpretation, then, involves the concept of gene-controlled reaction rates.

In connection with melanin-enzyme systems in the rabbit, Onslow found that recessive whites lack the enzyme necessary to convert chromogen into pigment, whereas dominant whites have this enzyme but in addition have an enzyme-inhibitor or anti-enzyme.

Reaction Rates.—The rate concept of character differentiation has been applied to many situations, particularly by Goldschmidt. As another example we may consider briefly the vestigial (vg) series of alleles in Drosophila, studied by Mohr, Goldschmidt and others. The various combinations of these alleles can be arranged in such a way that, phenotypically, they form a more or less continuous series from normal to a condition in which the wing is practically absent (Fig. 114). Goldschmidt has found that in the allelic combinations that result in intermediate and more nearly normal wing types, the wings develop at the same rate as do those of normal flies up to a certain time, and thereafter undergo a local disorganization and resorption that leads to the characteristic patterns. Thus the wings of certain combinations are normal at the time of pupation, but shortly afterward local retrogression sets in. One interpretation proposed by Goldschmidt assumes that a specific lytic substance, on reaching a certain threshold concentration, results in localized lysis of wing tissue. The rate of production of this substance is gene-controlled. In normal flies its concentration never reaches

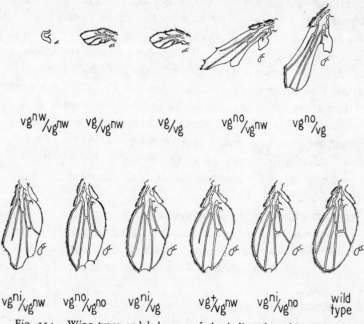

Fig. 114.—Wing types and balancers of the indicated combinations of vestigial alleles in Drosophila. Symbols: vg^{nw}, no-wing; vg^{no}, notch; vg^{ni}, nicked; vg, vestigial; and vg^+, wild-type. (After Mohr.)

the threshold level. In the various combinations of mutant genes, threshold concentrations are reached at various times in development. The earlier this concentration is reached the greater is the amount of tissue destruction and the smaller the final wing. Goldschmidt has also presented an alternative interpretation which assumes that the size of the wing is determined by the quantities of a hypothetical wing-forming substance present at various times of development.

HORMONES

Sex-hormones in Birds.—In both plants and animals specific diffusible chemical substances, known as hormones, regulate vital processes. In many instances they are known to be intermediaries

Fig. 115.—Male Sebright fowls. Above, normal hen-feathered male. Below, castrated male, cock-feathered.

between genes and characters. As an example—sex in dioecious animals is under control of the genetic system. This does not mean that the development of all characters related to sex is unalterably determined at the moment of fertilization. Under ordinary conditions the final sex is determined at fertilization, but no organism is completely resistant to modification by "extra-ordinary conditions." Secondary sex-characters in higher animals are largely dependent on so-called sex-hormones. In birds, for example, sex is normally determined by the chromosome con-stitution. But the production of sex-hormones, also normally under genetic control, can be altered experimentally. In some birds, of which the domestic fowl is an example, removal of the ovary (only the left ovary is functional in birds) from a potential female results in development of male-like plumage, *i.e.,* the female becomes "cock-feathered." The "hen-feathering" can be restored by injecting female sex-hormone or by implanting an ovary. Not only is hormone production here under genetic con-trol, but the reaction to hormone can likewise be shown to be under the influence of genes. In the Sebright bantam and some other breeds of fowl (*e.g.,* Campine), the males are normally hen-feathered. This hen-feathering of the male is inherited as a dominant in crosses with breeds in which the males are normally cock-feathered. From castration, gonad transplantation, and skin transplantation experiments, made by Morgan, Roxas, Danforth, and others, it can be shown that hen-feathering in the male of the Sebright and Campine is the result of a modification such that the feather follicles of the male react to male hormone in the same way as to female hormone (Fig. 115). As shown in Table 3,

TABLE 3

Type of feathering in males and females of hen-feathered breed of fowls (Sebrights or, in one case, Campines) and in a dimorphic breed (Leghorn) in relation to condition of gonad. From results of gonadectomy, gonad trans-plantation, or skin transplantation.

Breed and inherent sex	No gonad	Ovary	Testis
Leghorn female	cock	hen	cock
Leghorn male	cock	hen	cock*
Sebright female	cock	hen	hen
Sebright male	cock	hen	hen*

* Testis may be from Sebright or from Leghorn.

castration of either sex of any breed results in cock-feathering. In the normal breeds, female sex-hormone results in hen-feathering, male sex-hormone in cock-feathering, while in Sebrights or Campines either female or male hormone, whether derived from a normal breed or a hen-feathered breed, gives hen-feathering.

Dwarf Maize.—Another example in which reactions involving hormones are known to be gene-controlled is found in the comparison of normal and dwarf maize made by Van Overbeek. Nana, one of the several dwarf types known in maize, is differentiated from normal by a single recessive gene, *na*. Nana plants have very much shortened internodes, the mature plant being

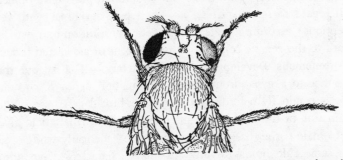

Fig. 116.—A gynandromorph of Drosophila melanogaster. Both front legs are male (sex-combs present). The left side of the head and thorax are wild-type and female; the right side eosin, forked, and male (except for a patch of wild-type tissue in the right eye). The specimen had two ovaries.

only about a foot or two in height. Nana plants apparently form the plant growth hormone, auxin, at the normal rate but oxidize it, *i.e.,* inactivate it, at a higher rate than normal. As a consequence the cells of a nana plant do not elongate to the same extent as do those of normals, and the entire plant is shorter. It appears, then, that the nana gene produces its characteristic phenotypic effect through a modification of the oxidation-reduction properties of the plant.

Gynandromorphs in Insects.—In the insects hormonal control of sex and secondary sex characters apparently does not occur, but instead these are controlled by intra-cellular factors. This is shown in a very simple way in individuals in which part of the body is XX in constitution and the remainder XY or XO. Such individuals, known as *gynandromorphs,* are mosaic for sex characters (Fig. 116). They result in two ways: (1) by elimination from one daughter cell at an early cleavage of one of the two X chromosomes (Morgan and Bridges), or (2) from double nucleus eggs (Doncaster). In the former, all descendants of the cell with a single X chromosome are genetically male while those from the sister XX cell are female. A double nucleus egg may or may not give rise to a gynandromorph, depending on whether the two nuclei are fertilized by like (X and X or Y and Y) or different (X and Y) sperms. Regardless of origin, gynandromorphs in Drosophila usually show autonomy of development with regard to sex-characters, *i.e.,* each part develops (with few exceptions) according to its own genetic constitution and without regard to the genetic constitution of adjacent or associated tissues.

Autonomous Development of Characters.—If a zygote that gives rise to a gynandromorph by elimination of an X chromosome is originally heterozygous for a sex-linked gene, say white, elimination of the X chromosome carrying the normal allele gives rise to male tissue that is at the same time genetically white-eyed. This same rule holds for any sex-linked character. Gynandromorphs can be obtained in which the male parts are forked and the female parts not-forked (Fig. 116). In such an individual the genetically forked parts show forked bristles. It is also found that forked parts and parts that show male sex characters always coincide so far as this can be determined. The forked character is, like secondary sex characters, autonomous (independent) in development, *i.e.,* shows no influence of the non-forked tissue in the same individual. This rule of autonomy of development applies to most characters differentiated by sex-linked genes. Vermilion is one exception to this rule to which we shall return later in this chapter.

Somatic Crossing Over in Drosophila.—There are other ways of obtaining individual organisms mosaic for tissues of differing genetic constitutions. We have already mentioned skin grafts in connection with feather characters in fowls. In many organisms

such grafts of parts originally from different individuals are technically easy to obtain and they are generally useful in studying
developmental processes. Stern has shown in Drosophila that
natural mosaics may occasionally arise as the result of *somatic
crossing over*. Such crossing over is normally very rare, but in
individuals of certain genetic constitutions (heterozygous for any
one of the several dominant minute genes—M_n, M_y, etc.) it occurs
with increased frequency. As a result of somatic crossing over
cells may be formed that are homozygous for genes which were
heterozygous in the original zygote. In individuals heterozygous
for two sex-linked genes such as *y* and *sn,* somatic crossing over

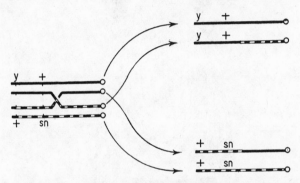

Fig. 117.—Scheme showing the way in which somatic crossing over may give
rise to cells homozygous for loci that were heterozygous before crossing over.

between the centromere and the singed locus (Fig. 117) may
give rise to daughter cells each homozygous for the two loci; one
for the recessive allele *y,* the other for the recessive allele *sn.*
If, instead of centromeres 1 and 3 going to the same pole, as
indicated in the diagram, 1 and 4 go to one pole and 2 and 3
to the other, the resulting cells will still be heterozygous for
both loci. If the *y/y* and *sn/sn* cells resulting from somatic
crossing over give rise to bristle-bearing surface tissue, so-called
"twin spots" will appear, one yellow and the other singed, on
a wild-type background (Fig. 118). The sizes of such twin
spots will of course depend on how early in development the
somatic crossover that gave rise to them occurred, on the rates of

growth of the daughter cells, and on the affected parts of the body. Usually they involve only a small portion of the surface parts of a fly. Somatic twin spots show that such characters as yellow and singed are autonomous in development when they occur in small amounts of tissue. By the use of the somatic crossover method, Demerec has studied the behavior of many sex-linked lethals in small patches of tissue. The general method,

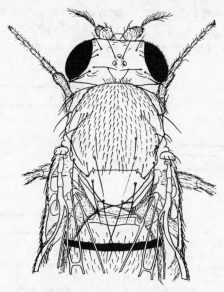

Fig. 118.—Twin spots due to somatic crossing over. Near the base of the right wing one bristle and ten hairs are singed; just to the left of these lie a bristle and seven hairs that are yellow (shown in the figure by dotted lines). (Drawn from data furnished by Stern.)

as illustrated in figure 119, involves the use of sex-linked non-lethal characters to mark the crossover twin spots. In the example illustrated in the figure, one of the twin spots is homozygous for y and a lethal and the other is homozygous for singed. If both spots regularly occur as twins, it is clear that in small patches of hypodermal tissue in a female the lethal does not prevent the survival and multiplication of cells homozygous for it. On the other hand, if only a single spot regularly occurs—singed in the illustration—it is concluded that cells homozygous for the lethal gene being studied do not survive. Demerec and Hoover have found that two types of lethals can be differentiated by this method; some are not lethal in small patches of female hypodermal tissue while others are lethal to such small groups of cells. In the latter case adjacent normal cells produce the tissue that the lethal-carrying cell normally would have produced. Most lethals that are demonstrated deficiencies for known loci are lethal in hypodermal tissue. There are two notable exceptions to this— deficiencies for yellow and for cut. Deficiencies for both of these loci are known which survive in homozygous condition in twin

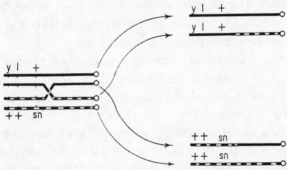

Fig. 119.—Somatic crossing over with yellow and singed used as indices of the presence of a lethal. Yellow areas are homozygous for the lethal; their presence in normal frequency (relative to singed spots) indicates survival of tissue homozygous for the lethal. Absence of yellow spots (when singed ones are numerous) indicates that tissue homozygous for the lethal does not survive.

spots. In the case of yellow deficiencies, tissue homozygous for the deficiency is phenotypically yellow. This is consistent with the fact, previously indicated, that a deficiency for the yellow gene may survive in the male, and is of interest in that it suggests that the most extreme known recessive allele of the yellow gene is an inactive form of the gene.

Lethals in Mice.—The above results show that a combination of genes lethal to an entire individual need not necessarily be lethal to individual parts of it. A similar conclusion was arrived at by de Aberle in studies of the dominant white mouse which is lethal in homozygous form. Homozygotes die of anemia soon after birth, but individual tissues transplanted to normal mice are able to survive. Somewhat comparable studies of the short-tailed mouse were made by Ephrussi. Short-tailed homozygotes die in the uterus, usually about 10 days after insemination, but if such embryos are removed shortly before the normal time of death, individual tissues can be removed and grown in standard type tissue cultures (in a mixture of embryo extract and blood plasma from the fowl). Several tissues have been so explanted and found to survive for periods of time far exceeding those over which the entire embryo would have lived. There appears to be no good reason why these tissues could not have been cultured indefinitely. Homozygous short-tailed embryos probably die as a result of some disturbance of general organization and not because of inviability of the component tissues.

Eye-color Hormone in Ephestia.—In the meal moth, Ephestia kühniella, there exists a mutant strain with red eyes instead of the black eyes characteristic of the wild-type. Caspari made the interesting observation that a testis from a wild-type larva (a^+/a^+) transplanted to a genetically red-eyed (a/a) larva, modifies development of the host in such a way that the genetically a/a adult has black rather than red eyes. The testis implant apparently produces a substance which diffuses into the eyes of the host and there modifies pigment development. Caspari, Kühn, Plagge, Becker, and others have since followed this lead and have found that the substance responsible, called the "a^+-hormone," is produced by ovary and brain tissue as well as by testes, and that it is concerned with several aspects of pigmentation—with larval skin-color, larval eye-color, testis-color, color of the brain, as well as with the eye-color of the imago. Here, then, a specific hormone is concerned in the development of a hereditary character. The presence or absence of the a^+-hormone depends on the genetic constitution of the individual with respect to the a gene. If, however, the hormone is absent because of the genetic constitution, the eye-color is not irreversibly determined; artificially supplying an a/a individual with the hormone results in the development of the eye-color characteristic of an a^+ individual.

In other words, a gene-conditioned developmental deficiency can in this particular instance be compensated for in much the same way as an insulin deficiency in a diabetic human being can be artificially compensated for by injections of the hormone insulin.

It should be emphasized that this modification in eye-color in an a/a individual is purely a somatic change—the genes are not modified. In spite of the extragenic nature of the modification, there may be a maternal influence on the next generation. Thus, if an a^+ testis is transplanted to an a female larva, the eyes of the host will be modified toward wild-type. Such a female, mated to an a/a male, gives offspring that are all genetically homozygous for the a allele. Nevertheless, they show pigmentation of the larval ocelli, which are normally colorless in a/a larvae. This is the result of an uptake of a^+-hormone by the eggs of the original host, and its transmission to the individuals that develop from these eggs. This is comparable to the transmission of a^+-hormone from an a^+/a female to a/a offspring. There is no corresponding transmission of hormone from an a^+/a male to an offspring. This example is one of the few cases in which a known substance is transmitted through the egg in the cytoplasm. This maternal influence lasts for only one generation.

Vermilion in Drosophila Mosaics.—As shown by Sturtevant, the vermilion character in Drosophila does not always show autonomous development in mosaics. In gynandromorphs in which large parts of the body are female and $+/v$ in constitution, eye-tissue that is genetically male and vermilion is often phenotypically not vermilion. That such tissue is really genetically vermilion can be shown by using a second eye-color character, such as garnet, known to be autonomous in gynandromorphs. As shown in figure 116, such a character can be used to "mark" the vermilion tissue. When the vermilion character is expressed, the v g eye-tissue is phenotypically v g (yellowish red) but, when the vermilion character is not autonomous, the v g eye-tissue is like g (dull pinkish) phenotypically. The presence of a body character such as forked is also useful in indicating the distribution of female and male tissues (Fig. 116).

Eye-color Hormones in Drosophila.—It appears, from this behavior of vermilion in mosaics, that normal eye-pigmentation in Drosophila must be dependent on a diffusible substance in somewhat the same way as in Ephestia. By implanting embryonic eye buds taken from vermilion larvae into the body cavities of wild-type larval hosts, Ephrussi and Beadle showed that the implant develops into an eye with wild-type pigmentation. The behavior of vermilion is essentially the same here as in mosaics. Various experiments have shown that the factor involved in such modifications of vermilion is a hormone-like substance. Similar transplantation studies with other eye-color characters in Drosophila have shown that the second chromosome recessive cinnabar, which is phenotypically indistinguishable from vermilion, is likewise non-autonomous. The phenotypic similarity of vermilion and cinnabar suggested that they might be similarly differentiated from wild-type, $i.\ e.,$ both might be deficient in the same hormone-like substance. This is shown not to be the case by the behavior of reciprocal transplants, which give the following results:

Constitution of transplanted eye bud	Host	Phenotype of implanted eye
vermilion cinnabar	cinnabar vermilion	wild-type cinnabar-like

The simplest interpretation of these results is that two different eye-color hormones are involved in the production of wild-type pigmentation; that a cinnabar fly is deficient in one of these, cn^+-hormone, and that a vermilion fly is deficient in both of them, v^+-hormone as well as cn^+-hormone. It is further assumed that these two eye-color substances are so related in development that cn^+-hormone is never formed in the absence of v^+-hormone. In a cinnabar fly, some one or more of the reactions connecting the two hormones is blocked and as a consequence v^+-hormone, but not cn^+-hormone, is formed. In the experiment in which a vermilion eye is grown in a cinnabar host, the host supplies the implant with v^+-hormone but not cn^+-hormone. Once the implant has a supply of v^+-hormone, there is no block to the formation, by the implant tissue itself, of cn^+ substance.

The above interpretation having to do with eye-color development in Drosophila has been tested in various ways by various investigators. It has been possible to extract the hormones from pupae and to purify them partially. Quantitative methods of working with these hormones have been developed, certain of their relations to pigment production are known, and some of their chemical characteristics have been determined.

Comparison of Hormones in Different Species.—The eye-color character ivory in Habrobracon, a wasp parasitic on Ephestia, is, like red eye in Ephestia and both vermilion and cinnabar in Drosophila, non-autonomous in mosaics (Whiting). As a result of cross comparisons involving Drosophila, Ephestia, and Habrobracon in which, for example, hormone extracts of Ephestia were tested in Drosophila for effects on vermilion and cinnabar, it has been possible to show that the hormones in these insects, which belong to three different orders, are apparently the same. The red eyed mutant in Ephestia is deficient in both v^+- and cn^+-substances, and therefore corresponds to the vermilion character in Drosophila. On the other hand, ivory in Habrobracon is deficient in cn^+-hormone but not in v^+-hormone, and therefore corresponds to cinnabar in Drosophila. It seems probable that the ivory (o^1) gene in Habrobracon and the cn gene in Drosophila are parallel mutations of homologous genes. In the same way, it seems reasonable to suppose that the a gene in Ephestia and the v gene in Drosophila are homologous. Whether or not these genes with corresponding effects are really homologous, the above comparisons shows that the eye-pigment systems must be essentially similar in the three orders of insects represented. Presumably, then, these are not genes that were important in differentiating these forms from each other during the process of evolution, but are representatives of the inheritance common to all of them—in other words, belong to the group of genes that make all three forms insects.

DIFFERENT WAYS OF PRODUCING THE SAME PHENOTYPE

The vermilion and cinnabar characters represent a situation frequently met with—that of two or more phenotypically similar characters differentiated from normal by different genes. The character scarlet eye-color in D. melanogaster is likewise phenotypically similar to vermilion and cinnabar. The three genes associated with these three characters modify development in different ways. Vermilion and cinnabar both have to do with eye-color hormones but, as pointed out above, they influence the chain of reactions at different levels. Scarlet flies, on the other hand, are characterized by the presence of both v^+- and cn^+-hormones, but these apparently cannot be made use of. The end results in the three characters, then, appear to be the same—the elimination of one pigment component in the eye. In each, this is brought about in a manner different from that of the other two. The phenotypic interactions of these three characters are consistent with the interpretation just given. For example, the double recessive cinnabar scarlet is phenotypically like either of the single recessives. The single recessives are each deficient in the same pigment component and the two together are consequently no different from either alone so far as the final result is concerned. A cross between cinnabar and scarlet gives a 9:7 ratio in F_2 as would be expected. Other examples of this same type of character interaction were discussed in Chapter III.

Environmental Modifications.—In many instances developmental effects similar to those brought about by known gene substitutions are produced by specific modifications of the environment. A familiar example of this kind is that of chlorophyll development in plants. It is known that any one of a dozen or more possible gene substitutions in maize results in a failure of chlorophyll development. There are evidently many different and genetically independent ways of blocking the reactions essential for the appearance of chlorophyll. The same end result—an albino plant—can be achieved in a very simple way by growing maize plants in the absence of light. Certain strains fail to develop chlorophyll at moderately low temperatures—temperatures at which other strains appear to be quite normal.   Another example of a strong modification of a character by temperature is found in the vestigial series of alleles in Drosophila.   Roberts, Harnly, and others have shown that the wing size of flies homozygous for the *vg* allele is greatly increased if flies are grown at 32° C. (25° C. is the usual temperature at which Drosophila experiments are made).   Figure 120 shows the relation of wing size to temperature in vestigial females which are also homozygous for a sex-linked modifier of vestigial (dimorphos).

Many other examples similar to those mentioned above could be given.   They all illustrate the fact that genes and environ-

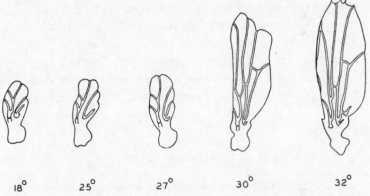

| 18° | 25° | 27° | 30° | 32° |

Fig. 120.—Effect of temperature on wing size in Drosophila males recessive for vestigial (*vg*) and a sex-linked specific modifier, dimorphos.   Below the outline drawings are indicated the temperature to which the flies were subjected during development.   (After Harnly and Harnly.)

mental factors are both components of the same system.   Both are essential parts of the system, and, in this sense, neither one nor the other can be said to be more important.

Non-correspondence of Characters.—In the above discussion, the fact that similar phenotypic effects can be brought about by different non-allelic genes is emphasized.   As indicated, this similarity in end results, *i.e.,* in phenotypes, does not at all mean a similarity in action of the various genes concerned.   In fact, in the case of the characters vermilion, cinnabar, and scarlet, it is known that the means of attaining the end result are different in each. It should be pointed out that correspondence, even of phenotypic effects of mutant genes, is by no means a general rule. As indicated in Chapter XX, it is possible to be reasonably sure of identifying homologous loci in different species of Drosophila where the only bases of comparison are phenotypic characteristics of the known mutant types. This is possible because, in general, mutants of a particular locus have phenotypic characteristics different from those of other loci. Often careful study of mutants that appear to be quite similar shows the existence of such characteristic differences. Thus in D. pseudoobscura there is a dominant character called *scute* because of its resemblance to scute in D. melanogaster. The resemblance is only superficial, however, since the pattern of bristle removal is different in the two types. The true homolog of scute in D. melanogaster is, with little doubt, the character known as scutellar. Even in the case of the phenotypically similar vermilion-like eye-colors, homologies can be established between species that cannot be crossed, by study of the developmental characteristics of the mutants—*i.e.*, by means of more complete descriptions of the characters.

CHEMICAL STUDIES

Anthocyanin Pigments.—Pigment systems in plants are favorable for studies of the developmental properties of hereditary characters. The pigments primarily responsible for flower colors in seed plants include: (1) yellow plastid pigments (carotenes and xanthophylls), (2) yellow anthoxanthins (flavones or flavonols), and the anthocyanins. The chemistry of all of these general groups of pigments is relatively well-known. Several investigators, particularly Scott-Moncrieff, have studied the flower pigments in various genetic types of several different plants. Among the many factors that affect pigment formation and distribution in flowers, the following are known to be gene-controlled:

1. Yellow plastid pigment production
2. Yellow anthoxanthin production

3. Ivory anthoxanthin co-pigment (substance having an effect on pigments out of proportion to its own color) production
4. General anthocyanin production
5. Specific anthocyanin production
6. Oxidation of pigment molecules
7. Oxidation and methylation of anthocyanins
8. Local changes in pH

Without extended discussion of these, two examples may be given. A gene pair (symbol *e*) is known in the Shirley poppy in which the dominant allele leads to the formation of antho-

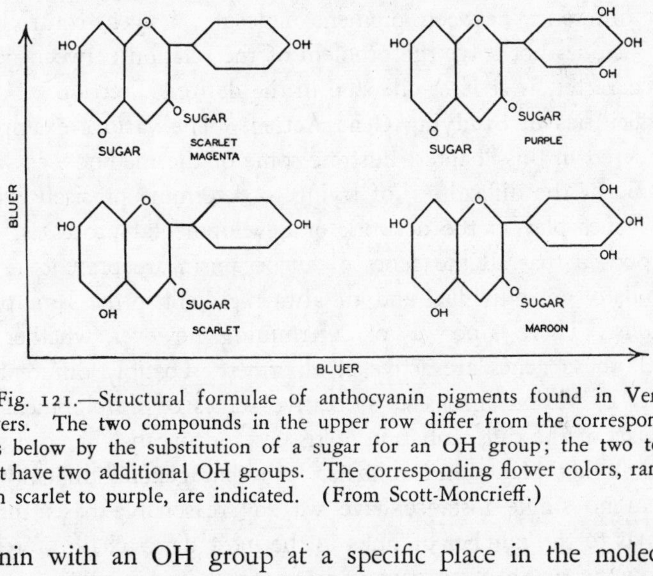

Fig. 121.—Structural formulae of anthocyanin pigments found in Verbena flowers. The two compounds in the upper row differ from the corresponding ones below by the substitution of a sugar for an OH group; the two to the right have two additional OH groups. The corresponding flower colors, ranging from scarlet to purple, are indicated. (From Scott-Moncrieff.)

cyanin with an OH group at a specific place in the molecule, whereas substitution of the recessive allele of this gene changes the pigment-forming reactions in such a way that this particular OH group is replaced by an H atom. With the oxidized form of the anthocyanin present the poppy flower is crimson with a black blotch at the base of each petal. The less oxidized form of the pigment gives a salmon flower with pale brown blotches. The chemical characteristics of the anthocyanin pigments of four genetic types of verbenas are shown in figure 121. Here both the number of OH groups and the number of sugars in the molecule vary. In Primula there is a gene pair (symbol r) that differentiates between low and high hydrogen ion concentrations (pH about 6.0 and 5.4 respectively) in the petals. In the presence of anthocyanins, which are hydrogen ion indicator pigments, the recessive form with the higher pH may be blue, while, on the same genetic background, the dominant form with the lower pH is magenta. This pH difference appears to be strictly localized in the flower petals.

In such cases as those discussed above it is known that reactions are gene-controlled, but it is not known how this control is brought about. The establishment of the fact that an oxygen atom difference between pigment molecules is gene-controlled, while it does not solve the problem of the relation between gene and character, is at least one step in the desired direction.

Difficulties in Studying Gene Action.—The various examples considered in this chapter illustrate some of the methods, as well as some of the difficulties, of trying to determine precisely what rôles genes play in the network of developmental processes. It is supposed that all the genes of an organism are present in all the cells of an individual and are always present in the same proportions. There is no way of determining, however, whether or not all these genes are active at all times. Possibly differential activity of various genes at successive stages of development is the basis of the differentiation of tissues and organs characteristic of all higher plants and animals. Another general difficulty is that in no single instance have we any reasonable basis for a guess as to the number of links in the most direct chain of reactions connecting a given gene and a given character. We usually assume that there are many, but this is not necessarily so. In such characters as those concerning the agglutination reactions that underlie the human blood groups, where there is no interaction between different alleles of a gene (*e.g.*, an A/B individual has both A and B agglutinogens and in this respect is the simple sum of A/A and B/B), it is possible that the number of steps between gene and agglutinogen is not great.

Looking to the future of genetics, it seems reasonably certain that its integration with such fields as those of biochemistry, developmental physiology, and experimental embryology will continue at an increasing rate. Regardless of whether or not studies designed to further such integration are likely, in the near future, to lead to an understanding of the nature and rôle of genes, there can be no question that they will add to our understanding of those processes that make up development.

REFERENCES

Ephrussi, B. 1938. Aspects of the physiology of gene action. Amer. Nat., 72:5–23.

Goldschmidt, R. 1938. Physiological genetics. 375 pp. McGraw-Hill Co., New York.

Kühn, A. 1937. Entwicklungsphysiologisch-genetische Ergebnisse an Ephestia kuehniella. Zeits. ind. Abst. Vererb., 73:419–455.

Morgan, T. H. 1934. Embryology and genetics. 258 pp. Columbia University Press, New York.

Muller, H. J. 1932. Further studies on the nature and causes of gene mutations. Proc. 6th Internat. Congr. Genetics, 1:213–255.

PROBLEMS

1. What would be expected following somatic crossing over between the loci of yellow and singed in a female of the constitution $\dfrac{+\ sn}{y\ +}$?

2. Using H for the symbol for hen-feathering in male fowls, how would you classify the hens in the F_2 of the cross Sebright by brown Leghorn with respect to their constitution for the H locus?

3. If transplants of wild-type larval fat bodies are made to vermilion larvae of Drosophila the adult hosts will be phenotypically wild-type. Similar transplants of wild-type larval salivary glands to cinnabar larvae do not modify the eye-color of the adult hosts. What would you conclude from these two experiments?

4. Assuming that a method is available by which it is possible to extract eye-color hormones from either Drosophila or Ephestia, what kinds of cross injection tests between these two forms would be necessary to show that the vermilion and a mutants in the two species are deficient in the same hormone (or physiologically interchangeable ones), and that cinnabar and a are not deficient in the same hormone?

CHAPTER XXIII

HISTORICAL

THE development of modern genetics began with Mendel's paper on peas—or, rather, with its rescue from oblivion in 1900. This paper was published in 1866, but made no impression on other biologists; when attention was called to it in 1900 its importance was at once recognized, and further development along the lines laid down by Mendel was immediate and rapid. Why the difference in reception?

One reason is probably that Mendel himself failed to confirm and extend his own results. After the pea experiments he undertook hybridization experiments with hawkweeds (Hieracium), and found that all his hybrids bred true. It is now known that this result is due to the fact that these plants usually reproduce by *apomixis—i.e.,* the embryo in the seed is formed directly from diploid maternal tissue, without the intervention of meiosis or fertilization. To Mendel, however, the result was an indication that the principles worked out for peas were not generally applicable. He therefore published no more accounts of his discoveries, and was thus partly responsible for their neglect—owing merely to an unlucky choice of material for his later work.

Developments before 1900.—In the period between 1866 and 1900 there were many developments in biology that made for a more ready appreciation of Mendel's work. The foundations of the modern knowledge of cytology were laid in this period. Beginning with the accurate description of fertilization and the tracing of the history of the sperm nucleus after it enters the egg (O. Hertwig in 1875, followed by Fol, Strasburger, and many others), there came the recognition of the chromosomes, of their multiplication by longitudinal splitting, and of their constancy in number (Flemming, van Beneden, Strasburger, and others,

1880–1885). By 1885 Hertwig and Strasburger had developed the conception of the nucleus as the basis of heredity.

Weismann (1885–1887, and later) formulated the germ-plasm theory, laying emphasis on the germ line as the conservative element in heredity, the successive individuals being produced by it but not themselves modifying it. This concept, the forerunner of the distinction between phenotype and genotype, led Weismann to deny the inheritance of acquired characters, and also paved the way for the appreciation of Mendel's factorial hypothesis. Weismann was also the first to lay emphasis on the theoretical implications of the constancy of chromosome number and of the facts of fertilization. He was led to predict that there must be a reduction in chromosome numbers at gametogenesis, as otherwise fertilization would result in a continuous increase in chromosome number. Boveri and others soon observed the occurrence of the reduction, and began the cytological study of meiosis. These results established the double nature of the somatic cells, as opposed to the single nature of the gametes—another of Mendel's postulates that must have seemed bizarre in 1866.

Still another development was that there was a growing interest in the study of discontinuous variation. Bateson in 1894, and deVries soon after, had begun a reaction from the view then current that all effective evolution is based on the selection of minute differences. It is significant that deVries was one of the three men to bring Mendel's paper to light, and that Bateson early became the most active of the Mendelian school.

Rediscovery of Mendel's Paper.—In 1900 there appeared three accounts, apparently all independent, by investigators (deVries, Correns, Tschermak) who had read Mendel's paper and appreciated its significance, and who were also able to confirm his principles from their own experiences. In the first of these papers deVries added eleven new cases that gave the 3:1 ratio in F_2, and these were found in ten widely different genera of seedplants. The generality of the principle was thus apparent at once. In 1902 the work of Cuénot and of Bateson showed that the same principles apply also to animals.

Chromosome Theory.—The close parallelism between Mendelian segregation and the behavior of the chromosomes was pointed out in 1902 independently by Sutton, Correns, and Boveri —the latter at the same time publishing his remarkable experiments on dispermic sea-urchin eggs, which furnished the first experimental proof that specific chromosomes differ in their effects on development, and that development is normal only in the presence of at least one complete set of chromosomes.

The chromosome theory of inheritance was next advanced by the discovery of the relation of the X chromosome to sex-determination. This was first suggested by McClung in 1902; but he failed to examine the chromosomes of the female, and supposed that she had no X instead of the two she actually possesses. In 1905 Stevens, and later Wilson, established the relations as we now know them. There was much activity in this field for several years; perhaps the most striking early work was that of Morgan on the chromosomes of the Phylloxerans (1908), in which it was shown that, even in a complex life-cycle which at first sight seemed wholly at variance with the sex-chromosome theory, in fact that theory gave the only intelligible interpretation of the cytological findings.

Linkage.—Linkage was first discovered, in the sweet pea, by Bateson and Punnett in 1906. The interpretation they gave is now discredited; but in the same year Lock suggested that, if homologous chromosomes undergo exchanges of materials (as had been suggested by Correns in 1902 on dubious theoretical grounds), then failure of such interchange might account for linkage—*i.e.,* he postulated that linkage is due to genes lying in a single chromosome pair, and that crossing over is due to exchange of materials between homologs. At about this time it began to be emphasized that there were, in some organisms, more pairs of genes than pairs of chromosomes. Since independent segregation was taken as a general law, and since most biologists were unwilling to admit exchange of material between chromosomes, it was argued that here was a fundamental objection to the chromosome interpretation of segregation. To Spillman belongs most of the credit for insisting that the argument was invalid, unless it could be shown that the number of *independently segregating* pairs of genes was greater than the number of pairs of chromosomes.

Cytological Basis of Crossing Over.—This argument seems to have influenced Janssens to look for evidence of exchange of material between chromosomes. His studies of meiosis, and particularly of the origin and behavior of chiasmata, led him in 1909 to publish an account which forms the starting point of the cytological study of crossing over.

Drosophila.—The next stage in the development of the chromosome theory was the result of the domestication of Drosophila, which was due to Woodworth and to Castle (1901). Several investigators used the material for studies on inbreeding and selection in the following years. In 1910 Morgan recorded several mutant types, one of which (white eyes) was sex-linked. New mutant types rapidly accumulated in the Columbia University laboratory; and Morgan's studies of these led to the proof that two pairs of sex-linked genes give recombination in the female (1910), and that this recombination is not a random one—*i.e.,* sex-linked genes are linked to each other in the female (1911). These results were interpreted as showing that homologous chromosomes do in fact exchange materials at meiosis. In 1912 Morgan correlated these results with Janssens' conclusions, and developed the general theory of linkage that still stands, according to which strength of linkage is inversely proportional to the likelihood that a crossover will occur. Sturtevant (1913) then elaborated these ideas by incorporating the conception of linear arrangement, and by constructing the first chromosome map. Double crossing over and interference were deductions that arose at once from this result.

Non-disjunction as Proof of the Chromosome Theory.—The phenomena of non-disjunction were studied by Bridges; in 1916 his full account of the genetics and cytology of non-disjunctional strains of Drosophila furnished final and convincing proof of the relation between genes and chromosomes. This paper also contained the first evidence indicating that crossing over occurs after the equational split—*i.e.*, in a four-strand stage; an alternative explanation was, however, possible until the work of Anderson (1925) on attached-X females.

Continuously Variable Characters.—The early Mendelian studies were, as a matter of convenience, based on discontinuous characters, since with these it is possible to make accurate classifications and to determine ratios. Such characters are still the basis of most critical work in genetics. A general theory of heredity, however, must take into account the nature and inheritance of continuous variability. For some years it was supposed that "blending inheritance" was different in kind from that described by Mendel, though Bateson, as early as 1902, had formulated the multiple gene interpretation for such cases. He pointed out that such a character as stature in man must depend on many developmental factors, and the existence of different allelic pairs affecting these would give a continuously variable population. This purely hypothetical formulation seemed unconvincing; but the experimental work of Nilsson-Ehle (1908–1909), East (1910) and Emerson (1910) furnished a solid basis of fact, on which was developed the theory as it now exists. With this development, a great portion of the field of genetics, especially on the practical side, became amenable to exact analysis.

Statistical Methods.—The analysis depends in large part on the use of statistical methods. These methods, developed chiefly by Galton and Pearson, were at first used in attempts to discredit the Mendelian interpretation. The recognition that statistical methods are, in fact, valuable working tools in Mendelian analysis came gradually. Perhaps the most significant single event here was the publication of Johannsen's "Elemente der exakten Erblichkeitslehre" (1909). This book not only emphasized the use of statistical methods; it also marked the beginning of the clear distinction between genotype and phenotype. These and several others of the standard terms of genetics (*e.g.*, "gene") are due to Johannsen.

Mutation.—The concept of mutation is, of course, old. Darwin was familiar with the occurrence of "sports," as were others long before him. It was Bateson and deVries who first laid emphasis on the importance of their study. The term mutation is due to deVries, who developed a general theory of evolution based on this phenomenon (1900). Unfortunately most of deVries' observations and experiments were made on Oenothera; as a result of the unusual nature of this plant, it happens that a large portion of the "mutations" on which his theories were based would not now be so described. Nevertheless, his essential idea of sudden and discrete changes in the nature of the germ plasm was correct, and greatly influenced the history of genetics. Only with the increasing knowledge of the genes and their behavior was it possible to obtain clear ideas about mutations. The early work of Morgan (1910–1912) on Drosophila, and the discovery of a specific mutable gene by Emerson (1917) in maize are perhaps the most significant contributions of the period in this field. In 1927 the work of Muller and of Stadler showed that mutation frequency can be greatly increased by irradiation; the study of mutation at once entered a new phase, which is still continuing.

Evolution.—Much of the early work in genetics was stimulated by an interest in evolution, for which an understanding of the nature of heredity is essential. It is clear that this is what led such men as Bateson and deVries to study heredity. The rapid development of the subject after 1900 led most geneticists to neglect the evolutionary implications—there were too many other things to be done. Evolution was not forgotten, but it seemed that more immediate and approachable problems were more profitable to attack. The properties of populations heterogeneous for allelic genes, however, lend themselves readily to mathematical analysis. It is partly through this fact that evolution has come back into style in genetics. Hardy worked out, in 1908, the equilibrium formula for a population heterogeneous for a single pair of alleles. Jennings (1916) studied the consequences of inbreeding, which were elaborated further by Wright (1921). Haldane began in 1924 a systematic mathematical study of the effects of selection on mixed populations. Fisher (since 1928) and Wright (since 1929) have contributed to the development of a detailed mathematical analysis of the effects of mutation, selection, and inbreeding.

Another line of study that has served to bring the attention of geneticists back to evolutionary problems is that of the cytology—and genetics—of species hybrids. This development was at first largely a botanical one, and may be dated, more or less arbitrarily, from the work of Rosenberg (1909) on Drosera. Important later contributions were those of Federley (1913) on the diploid sperm produced by certain hybrid moths, and of Winge, who developed the interpretation of allopolyploidy in 1917.

Interpretation of Oenothera Genetics.—Related to these studies was the cytological work of Belling on Datura, which led him in 1926 to the hypothesis of reciprocal translocations as an interpretation of chromosome rings. This view, applied to the cytological pictures already found in Oenothera by Cleland (1922), led to the final clearing up of the puzzling behavior of that plant —a clearing up already begun by Renner's demonstration in 1917 of the balanced lethal system in Oe. lamarckiana.

Cytological Advances.—Cytological study of the chromosomes is older than the study of Mendelian heredity. Its development has been more or less apart from that of genetics, and it remained largely a descriptive subject until the work of Darlington, whose book in 1932 may be taken as marking the unification of chromosome cytology. From then on it has been possible to see the subject as a whole, united by a series of working hypotheses.

Salivary Gland Chromosomes.—The most striking recent development in cytology is the discovery of the chromosomal nature of the long-known giant structures in the nuclei of the salivary glands of Diptera. This discovery, made with Bibio (Heitz and Bauer, 1933), was extended to Drosophila by Painter in 1934. The metaphase chromosomes of Drosophila are extremely small, and almost no structural details can be made out in them. This discovery, then, made Drosophila one of the best objects for the study of structural detail of chromosomes, whereas it had previously been one of the poorest. The resulting impetus to the development of genetics can not yet be fully appreciated.

Other Advances.—These are some of the events that seem to us important in the history of genetics. Much has been omitted— we might have included the newer studies on the determination of sex, on polyploidy, on the developmental effects of genes, or various other topics. Even in the fields discussed, we can only pretend to have mentioned a few of the more significant contributions. Our object has been to show something of the order in which ideas developed.

We should like to conclude the account with an outline of the future of genetics; but one of the things that makes science interesting is that new developments are likely to come in unexpected directions. One thing can safely be predicted; genetics will continue to develop and to include new ideas and new fields of work —it is not static.

REFERENCES

Bateson, B. 1928. William Bateson, naturalist. 473 pp. University Press, Cambridge.

Bateson, W. 1909. Mendel's principles of heredity. 396 pp. University Press, Cambridge.

Cook, R. 1937. A chronology of genetics. U. S. Dept. Agric. Yearbook, 1457–1477.

Iltis, H. 1932. Life of Mendel (translated by E. and C. Paul). New York.

Roberts, H. F. 1929. Plant hybridization before Mendel. 374 pp. University Press, Princeton.

Wilson, E. B. 1925. The cell in development and heredity. 1,232 pp. The Macmillan Co., New York.

APPENDIX

PROBABILITY

As indicated in Chapter I and repeatedly thereafter in this book, genetics is to a large extent a science of ratios. In all ratios of the kind dealt with by the geneticist sampling errors are of importance, and to appreciate the significance of experimentally determined proportions of different phenotypes, one must have a clear understanding of simple probability as it relates to such proportions.

It is suggested that a review of the sections on probability in any standard elementary algebra book will be of help to those students whose knowledge in this field may have suffered through disuse. An understanding of the simple laws of probability in addition to an understanding of the material given in this section, should enable the average student to appreciate the handling and interpretation of genetic data. It is urged, however, that those who anticipate following genetics beyond the limits of this book would do well to make use of one or more of the references given at the end of this appendix. Conscientious working of the problems to be found at the ends of the various chapters of this book cannot be too strongly urged.

The method of obtaining probabilities by the use of the binomial expansion $(p + q)^n$, was introduced in connection with the distribution of sexes in human families (Chapter I). Continuing with this same method, let us consider populations of six individuals where a 1:1 ratio is expected, say a^+ to a phenotypes from the cross $a^+/a \times a$. The probability of all individuals of a given sample of six being a^+ is of course $(\frac{1}{2})^6 = \frac{1}{64}$. Or, expressed as a decimal fraction, the probability that a given sample of six will consist entirely of a^+ individuals is 0.0156. Since all possibilities must give a total probability of 1.0 (*i.e.,* certainty), the probability that a given sample of six will *not* consist entirely of $a+$ individuals is 1.0 − 0.0156 = 0.9844. In other words, where the true ratio of $a+$ to a is 1:1, one could expect, on the average, to obtain 156 samples of 6 $a+$ individuals out of every 10,000 such samples, or, conversely, 9844 samples of 6 out of 10,000 in which not all individuals are $a+$. The probabilities of obtaining the various possible combinations of $a+$ and a individuals in a sample of six is given by the expression $(p + q)^6$ where p is the probability of $a+$ and q the probability of a. Expanded, this gives:

$$p^6 + 6\,p^5q + 15\,p^4q^2 + 20\,p^3q^3 + 15\,p^2q^4 + 6\,pq^5 + q^6$$

The terms can be taken to represent the probability of obtaining the various combinations of $a+$ and a, *i.e.,* p^5q signifies 5 $a+$ to 1 a individuals, and the coefficient, 6, tells us that there are six chances in 64 of obtaining this combination. Where a 1:1 ratio is expected, p and q are equal to each other and, since their sum must be 1, they are each equal to ½. Each term of the expression becomes $(½)^6$, and the coefficients therefore represent the relative frequencies with which the various combinations are expected. Thus, one expects 3 $a+$ to 3 a, on the average in 20 out of every 64 samples of six individuals, or 20 times as often as 6 $a+$ to 0 a. If the various terms are multiplied by their coefficients, the probability of each of the various combinations is obtained. These may be expressed either as fractions or decimals, as follows:

$$
\begin{array}{rcl}
6 + : 0\ a & = & \tfrac{1}{64} = 0.0156 \\
5 + : 1\ a & = & \tfrac{6}{64} = 0.0938 \\
4 + : 2\ a & = & \tfrac{15}{64} = 0.2344 \\
3 + : 3\ a & = & \tfrac{20}{64} = 0.3125 \\
2 + : 4\ a & = & \tfrac{15}{64} = 0.2344 \\
1 + : 5\ a & = & \tfrac{6}{64} = 0.0938 \\
0 + : 6\ a & = & \tfrac{1}{64} = 0.0156 \\
\end{array}
$$

Since the expression $(½ + ½)^6$ is 1^6, or 1, the total of the probabilities must be 1.0. The combinations 3 + to 3 a is a perfect fit to the expected ratio of 1:1. All other combinations deviate from this. Thus the class 6 + to 0 *a* deviates by three individuals, *i.e.,* three individuals would have to be changed to get a perfect ratio.

Writing down in terms of the *a* individuals only, the probabilities of getting the various possible deviations from a 1:1 ratio of *a*+ to *a*, we have:

Combination	6:0	5:1	4:2	3:3	2:4	1:5	0:6
Deviations	− 3	− 2	− 1	0	+ 1	+ 2	+ 3
Probability	.0156	.0938	.2344	.3125	.2344	.0938	.0156

This can be represented graphically as in figure 122, in which deviations are plotted on the horizontal axis and the probabilities of such deviations are represented by vertical lines. All of these

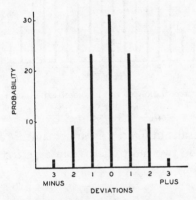

Fig. 122.—Graphic representation of the binomial expansion $(a + b)^6$, where $a = b = 0.5$. The heights of the columns are proportional to probabilities and the sum of their heights is therefore unity.

lines add up to 1.0 since their lengths represent probabilities. From the values given it is seen that the probability of getting 3 *a*+ to 3 *a* individuals in a given sample of 6 is 0.3125. In other words, the probability is 0.6875 that a random sample of 6 will not be made up of 3 of each phenotype—out of every 100 samples of 6, one would expect about 69 with either more or less than 3 *a* individuals.

24

If, instead of samples of 6 individuals, we consider groups of 50, we can determine probabilities in exactly the same way. The only difficulty is that it becomes somewhat laborious to expand the binomial $(p + q)^{50}$. However, if this is done and the results are expressed graphically in the same way as illustrated with 6

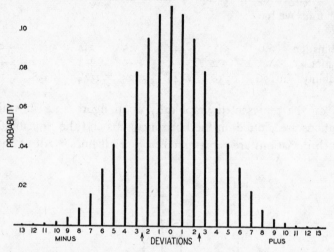

Fig. 123.—Graphic representation of the binomial expansion $(a + b)^{50}$, where $a = b = 0.5$.

individuals, the results in figure 123 are obtained. For deviations up to and including two individuals from a perfect fit of 25:25, the probabilities are:

— 2	0.0960
— 1	0.1080
0	0.1123
— 1	0.1080
— 2	0.0960

The sum of these is 0.5203, which is the probability of getting, in a sample of 50 individuals, of which two kinds are expected in equal numbers, a deviation of two or less from a perfect fit. Calling this value exactly 0.50 for simplicity, this means that in such a sample of 50 the chances of getting a deviation, from a 1:1, of more than two is equal to that of getting a deviation of two or less. These probabilities are of use to us in several ways. As an example, one might suspect a character Z in dogs of being inherited as a simple Mendelian dominant. From test crosses of $Z/+ \times +$ one might obtain 15 Z to 35 $+$ dogs in a population of 50. This represents a deviation of 10 from the expected 25 Z dogs. By working out the probabilities for various deviations as indicated in figure 123, one finds that the probability of obtaining, through errors of sampling, a deviation of 10 or more in either direction (*i.e.*, either too many or too few Z dogs) in a population of 50, is only 0.0064. That is, there are only about 6 chances in 1,000 of getting a deviation as great as this by chance if the true ratio is 1:1. This means that this particular population of 50 dogs very probably does not represent a random sample from a population in which the true ratio is 1 Z to 1 $+$. However, the deviation may indicate one of several things, for example:

1. The hypothesis of a simple dominant may not be correct:
2. The hypothesis may be correct, but Z dogs may be less viable than $+$ dogs.
3. The hypothesis may be correct and both types of dogs may be equally viable; in other words, the deviation is not *impossible* on the basis of chance but is merely *improbable*. Of a thousand samples of 50 dogs segregating in a true 1:1 ratio, about 6 would be expected to give a deviation as great as or greater than the observed one.

This analysis in terms of probability does not give the answer to the question: How is the Z character inherited? It does tell one that it is *very improbable* that Z is inherited as a simple dominant with no complications. Furthermore, it enables one to say just how improbable this is. The only satisfactory way to answer the question is to make more breeding tests. Even if the observed ratio had been 25:25, it would not be established that Z is inherited as a simple dominant—there would merely be a very good agreement between hypothesis and observation.

Where the number of individuals involved is large, this method of working out probabilities from the binomial expansion is not used because of the labor involved. The method is presented to illustrate the principle of a technique that is commonly used. If a probability diagram such as that for a 1:1 ratio with 50 individuals is made for a much larger number of individuals, say 1,000, the possible combinations will be much greater than for 50, and the number of vertical lines will be correspondingly increased. If the number of individuals is still further increased, the diagram will approach, as a limit, a solid figure with a smooth boundary (Fig. 124). Such a curve is known as a *normal curve*,

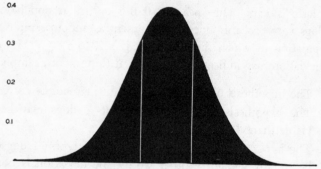

Fig. 124.—Normal curve such as is approached as a limit by the binomial $(\frac{1}{2} + \frac{1}{2})^n$. White lines divide the area under the curve in such a way that half lies between them and half beyond them.

or *curve of errors*. The vertical lines of figure 123 may be thought of as being fused and the area under the curve as being equal to the sum of the lengths of all of the lines. The area under the curve will then represent probability and the total area will therefore be unity. If such a curve be taken to represent deviations from a 1:1 ratio for n individuals, it is possible to calculate, from the mathematical characteristics of the curve, a deviation which, taken in both plus and minus directions will include exactly half the area under the curve, as shown in figure 124. This value, expressed in this case as a deviation, will vary with n and will be of such magnitude that a given sample of n individuals will be just as likely to deviate from a true 1:1 ratio by less than this value as by more than this value. Such a value is known as the *probable error*. As an example of this, the probable error for a 1:1 ratio where n is 880 is 10. This means that a sample of 880 individuals taken at random from an infinite population in which there are equal numbers of two types will be as likely to give between 450:430 and 430:450 as to give a deviation greater than 10. Thus, if one were to obtain an actual count of 430 to 450 where a 1:1 ratio is expected by theory, the agreement would be good. One would expect, through sampling errors alone, to get an agreement as bad or worse than this in one half of a large number of trials.

Probable errors can be calculated for any ratio by means of the simple formula, probable error $= 0.67449 \sqrt{p \cdot q \cdot n}$, where p and q are the theoretically expected proportions of the two types of individuals ($p + q$ must equal 1), and n is the number of individuals involved. Thus, in the above example for a 1:1 ratio with 880 individuals, the probable error is calculated as follows:

$$0.67449 \sqrt{0.5 \cdot 0.5 \cdot 880} = 0.67449 \sqrt{220} = 10$$

It should be pointed out that the application of this or other formulae for the calculation of probable errors is based on the assumption that frequencies of deviations are distributed in the form of a normal curve. The distribution of deviations for a 3:1 ratio will not be a normal one, but as shown by the expansion of $(p + q)^{50}$ where p is equal to ¾ and q is equal to ¼, will be asymmetrical, technically *skewed*. Strictly speaking, the use of the indicated formulae are not applicable here, but in practice the error is small unless the skewness is very strong. For 15:1 or 63:1 ratios, however, the error is large enough so that the use of the above formulae might be misleading (Fig. 125). Special methods have been developed for such cases, but these are beyond the scope of this book (see reference to Fisher). Returning to figure 123, which gives the probabilities of various combinations for a 1:1 ratio with 50 individuals, it can be seen by inspection that the *probable error* is approximately 2.5, *i.e.*, from the center to half way between 2 and 3. If, however, one calculates the probable error from the above formula a value of 2.38 $(0.67449 \sqrt{0.5 \cdot 0.5 \cdot 50} = 2.38)$ is found. This discrepancy results from the fact that the probability in the binomial diagram out to and including a deviation of 2 is not exactly 0.5 but 0.5203. Furthermore, the formula for the probable error is based on the normal curve while the binomial expansion, where n is 50, is only an approximation to this. It follows that for small numbers, say 25 or less, it is more accurate to use the binomial method of calculating probabilities. It is not necessary to work these out

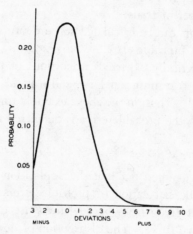

Fig. 125.—Curve obtained from the binomial expansion $(a + b)^{48}$, where $a = \frac{15}{16}$ and $b = \frac{1}{16}$; *i.e.*, for a 15:1 ratio of a to b.

as they are available in the form of tables for all values of n up to 50 (Warwick—see References).

It is not necessary to use probable errors in the form of numbers —it is a simple matter to express them in terms of percentages. Thus in the diagram of probabilities for 50 individuals (Fig. 123), a deviation of one individual is equal to a deviation of two per cent, *i.e.*, the numerical ratio is 26:24 or 24:26 which, expressed in terms of percentages, is 52:48 or 48:52—a deviation of two per cent from the expected 50 per cent. In terms of percentages the probable error would then be 4. The general formula for calculating probable errors in percentages is

$0.67449 \sqrt{\dfrac{p \cdot q}{n}} \times 100$ where p, q, and n have the same significance

as in the formula for the probable error in numbers.

Depending on the values of the constants in the equation for it, a normal curve may be broad or narrow (Fig. 126). Regardless of its shape the area included within a range, extending from the

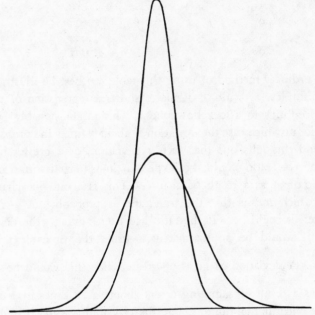

Fig. 126.—Two normal curves indicating a difference in variability. Both curves are of unit area.

center, to a given number of times the probable error is constant. This means that the probability of a deviation of any given number of times the probable error can be calculated. Since these probabilities are the same for all normal curves it is a simple matter to calculate them once and arrange them in a table which can be consulted at will. Thus, the probabilities, *i.e.*, areas under the curve, outside various ranges of the center are as follows:

Deviations in probable errors, i.e., $\dfrac{\text{deviation}}{\text{probable error}}$	Probability of obtaining deviation as great or greater by chance
1	.500
1.5	.312
2.0	.177
2.5	.092
3.0	.043
3.5	.018
4.0	.007
4.5	.002
5.0	< .001

The ordinary method of using this table can best be illustrated with examples. A culture of maize shows a segregation of 220 green seedlings to 180 yellow ones. This might possibly be a 1:1 ratio, in which case the deviation is about 3 times the probable error and this tells one that such a deviation, or a greater one, from a 1:1 ratio could be expected in 43 trials in 1,000 (p = 0.043) as a result of chance. This is about one chance in 23 which means that the true ratio is probably not a 1:1. However, it could be. If the ratio had been 227:173 the deviation, 27, would be approximately 4 times the probable error $\left(\dfrac{27}{6.7}\right)$ A deviation as great or greater than this would be expected as a result of sampling errors about seven times in 1,000 trials or once in 141 times. Here the probability is much higher that the true ratio is not 1:1. In the first example, 220:180, green plants might be differentiated from yellow by two dominant complementary genes, and the expected ratio of green to yellow (in F_2) therefore be 9:7. The expectation on this basis is 225:175, from which the observed ratio 220:180 deviates by only 5 individuals. The probable error for a 9:7 ratio where n is 400 is 6.7 (the same, to the first decimal, as for a 1:1 ratio). Since the deviation is less than the probable error the true ratio *could be* 9:7 from the statistical standpoint. However, this does not prove that it is 9:7—further experimental data are needed to do this.

Furthermore, even in the assumed example of 227:173, where the probable error tells us that the deviation is probably not a chance one from a true 1:1 ratio, the data do not exclude a one factor difference—the observed ratio may deviate from 1:1 because the seeds that give yellow seedlings do not germinate as well as do those that give green seedlings. Again, further experimental data are needed to solve the problem of how the yellow seedling character is inherited.

The above examples illustrate that all statistical methods can tell one about ratios is what deviations one can reasonably expect as a result of sampling errors. They may indicate that a given hypothesis does not agree with observed facts, but they do not tell one in what respect the hypothesis is faulty. A hypothesis may be wrong in assuming that one gene is segregating instead of two, or merely in that it does not take into account differential viability—a calculation of probable errors does not necessarily distinguish between two such alternatives.

Although the calculation of probable errors is simple, tables are available for many standard genetic ratios, and values for probable errors can be looked up in these (Castle, see references).

Probable errors can be applied to crossover values as well as to ratios. Since these are usually expressed on a percentage basis, the formula is

$$\text{Probable error} = 100 \times 0.67449 \sqrt{\frac{p \cdot q}{n}} \quad \text{where } p = \text{proportion}$$

of crossovers, and $q = 1 - p$. As in the case of Mendelian ratios, the above formula for the probable error of a percentage is entirely justified when deviations from a mean percentage are symmetrically distributed. For very low percentages, the formula should be used with caution.

In the formulae given for calculating probable errors a factor, 0.67449, is present. The value obtained without this factor is known as the *standard error* (abbreviated S.E.). Since the two values, probable error and standard error, always bear a constant relation to each other as indicated by the factor 0.67449, there is no particular point in going to the extra work of getting probable errors. Both standard error and probable error are used in genetic literature and one should be careful to note which one is used. For converting deviations of various numbers of times the standard error into probability values, a different table is needed than that used for probable errors. A few values are given as follows:

Deviation / Standard error	Probability of obtaining a deviation as great or greater by chance	Odds against (to 1)
1.0	.3174	2.15
1.5	.1336	6.5
2.0	.0456	21
2.5	.0124	80
3.0	.0026	380
3.5	.0004	2,500
4.0	< .0001	17,000

In this table there is a column headed "Odds against (to 1)." This is simply another way of expressing the probability of a deviation of more than a certain magnitude. It is the ratio of the probability of obtaining a deviation of less than a given magnitude (expressed in standard error units) to that of obtaining one of more than the specified magnitude. Odds against of 21 means that there are 21 chances to 1 against getting a deviation as great or greater than the one specified through sampling errors alone. Two times the standard error is equivalent to odds against of 21 and this is frequently taken as an arbitrary level of significance, *i.e.*, a deviation of less is considered not statistically significant and one or more as significant from the statistical standpoint. Obviously common sense should be used in applying any such concept.

The theory of probability is as applicable to measurements as to ratios of various kinds. Thus, in the example given in Chapter XVII of the inheritance of ear-length in maize, measurements of ears from individual plants were made for both parents, for the F_1 plants and for the F_2 plants. Considering only the F_1 plants, the distribution of lengths of individual ears is shown in figure

101 to be roughly that of a normal curve. From the observed distribution it is possible to calculate the standard error (or probable error if this is desired) of the mean (arithmetic average) ear-length. In the example chosen the mean ear-length in F_1 is 12.6 cm. The standard error of this is 0.19 cm. The mean is usually written together with its standard error in the following way: 12.6 ± 0.19. This standard error of the mean gives a measure of the probability of deviations from the true mean of a given magnitude occurring as a result of random sampling. Thus, there are about three chances of the measured mean deviating from the true mean by the amount of the standard error to two chances of its deviating by less than the standard error. In the example chosen the mean ear-length of the F_1 plants is 12.6 ± 0.19, which indicates that the true mean is probably within 0.38 (2 times the standard error) of that found.

It is a simple matter to determine the significance of a difference between two measured values. The mean ear-length of F_1 maize plants is 12.6 ± 0.19 cm., while that of the long-eared parent is 16.8 ± 0.18 cm. The difference of these two means is 4.2 cm. Some difference could be expected due to errors of sampling involved in the determination of both means; the standard error of the difference tells one just how much could reasonably be expected. The standard error of a difference between two non-correlated values is found from the formula

$$\text{S.E.}_{\text{diff.}} = \sqrt{(\text{S.E.}_1)^2 + (\text{S.E.}_2)^2}$$

where S.E.$_1$ is the standard error of one mean and S.E.$_2$ the standard error of the other. Using this formula, the standard error of the difference in the two ear-length means given above is found to be 0.26. The difference is thus 4.2 ± 0.26 or about 16 times the standard error. The probability of such a difference being the result of errors of sampling is almost zero, much less than 0.0001.

Here again the use of the usual formulae for standard errors assumes a normal distribution of measurements; this fact should be constantly kept in mind.

Methods of calculating standard errors of the means of measured values can be found in any one of the several books referred to below. Other useful statistical methods such as those for determining agreement between observed and expected ratios where there are more than two phenotypes, will likewise be found in these references.

REFERENCES

Castle, W. E. 1924. Outline for a laboratory course in genetics. Harvard University Press, Cambridge. (Contains Emerson's Probable Error tables.)

Davenport, C. B., and M. P. Ekas. 1936. Statistical methods in biology, medicine, and psychology. John Wiley and Sons, New York.

Fisher, R. A. 1936. Statistical methods for research workers. 6th Ed. 339 pp. Oliver and Boyd, Edinburgh.

Mather, K. 1938. The measurement of linkage in heredity. 132 pp. Methuen and Co., London.

Sinnott, E. W., and L. C. Dunn. 1932. Principles of genetics. 441 pp. McGraw-Hill Co., New York. (Appendix by D. R. Charles on Biometric methods.)

Warwick, B. L. 1932. Probability tables for Mendelian ratios with small numbers. Texas Agric. Exper. Station. Bulletin 463.

INDEX

A CATALOGUE OF SELECTED DOVER BOOKS
IN ALL FIELDS OF INTEREST

A CATALOGUE OF SELECTED DOVER BOOKS
IN ALL FIELDS OF INTEREST

AMERICA'S OLD MASTERS, James T. Flexner. Four men emerged unexpectedly from provincial 18th century America to leadership in European art: Benjamin West, J. S. Copley, C. R. Peale, Gilbert Stuart. Brilliant coverage of lives and contributions. Revised, 1967 edition. 69 plates. 365pp. of text.

21806-6 Paperbound $3.00

FIRST FLOWERS OF OUR WILDERNESS: AMERICAN PAINTING, THE COLONIAL PERIOD, James T. Flexner. Painters, and regional painting traditions from earliest Colonial times up to the emergence of Copley, West and Peale Sr., Foster, Gustavus Hesselius, Feke, John Smibert and many anonymous painters in the primitive manner. Engaging presentation, with 162 illustrations. xxii + 368pp.

22180-6 Paperbound $3.50

THE LIGHT OF DISTANT SKIES: AMERICAN PAINTING, 1760-1835, James T. Flexner. The great generation of early American painters goes to Europe to learn and to teach: West, Copley, Gilbert Stuart and others. Allston, Trumbull, Morse; also contemporary American painters—primitives, derivatives, academics—who remained in America. 102 illustrations. xiii + 306pp.

22179-2 Paperbound $3.50

A HISTORY OF THE RISE AND PROGRESS OF THE ARTS OF DESIGN IN THE UNITED STATES, William Dunlap. Much the richest mine of information on early American painters, sculptors, architects, engravers, miniaturists, etc. The only source of information for scores of artists, the major primary source for many others. Unabridged reprint of rare original 1834 edition, with new introduction by James T. Flexner, and 394 new illustrations. 6⅝ x 9⅝.

21695-0, 21696-9, 21697-7 Three volumes, Paperbound $15.00

EPOCHS OF CHINESE AND JAPANESE ART, Ernest F. Fenollosa. From primitive Chinese art to the 20th century, thorough history, explanation of every important art period and form, including Japanese woodcuts; main stress on China and Japan, but Tibet, Korea also included. Still unexcelled for its detailed, rich coverage of cultural background, aesthetic elements, diffusion studies, particularly of the historical period. 2nd, 1913 edition. 242 illustrations. lii + 439pp. of text.

20364-6, 20365-4 Two volumes, Paperbound $6.00

THE GENTLE ART OF MAKING ENEMIES, James A. M. Whistler. Greatest wit of his day deflates Oscar Wilde, Ruskin, Swinburne; strikes back at inane critics, exhibitions, art journalism; aesthetics of impressionist revolution in most striking form. Highly readable classic by great painter. Reproduction of edition designed by Whistler. Introduction by Alfred Werner. xxxvi + 334pp.

21875-9 Paperbound $3.00

JIM WHITEWOLF: THE LIFE OF A KIOWA APACHE INDIAN, Charles S. Brant, editor. Spans transition between native life and acculturation period, 1880 on. Kiowa culture, personal life pattern, religion and the supernatural, the Ghost Dance, breakdown in the White Man's world, similar material. 1 map. xii + 144pp.
22015-X Paperbound $1.75

THE NATIVE TRIBES OF CENTRAL AUSTRALIA, Baldwin Spencer and F. J. Gillen. Basic book in anthropology, devoted to full coverage of the Arunta and Warramunga tribes; the source for knowledge about kinship systems, material and social culture, religion, etc. Still unsurpassed. 121 photographs, 89 drawings. xviii + 669pp.
21775-2 Paperbound $5.00

MALAY MAGIC, Walter W. Skeat. Classic (1900); still the definitive work on the folklore and popular religion of the Malay peninsula. Describes marriage rites, birth spirits and ceremonies, medicine, dances, games, war and weapons, etc. Extensive quotes from original sources, many magic charms translated into English. 35 illustrations. Preface by Charles Otto Blagden. xxiv + 685pp.
21760-4 Paperbound $4.00

HEAVENS ON EARTH: UTOPIAN COMMUNITIES IN AMERICA, 1680-1880, Mark Holloway. The finest nontechnical account of American utopias, from the early Woman in the Wilderness, Ephrata, Rappites to the enormous mid 19th-century efflorescence; Shakers, New Harmony, Equity Stores, Fourier's Phalanxes, Oneida, Amana, Fruitlands, etc. "Entertaining and very instructive." *Times Literary Supplement.* 15 illustrations. 246pp.
21593-8 Paperbound $2.00

LONDON LABOUR AND THE LONDON POOR, Henry Mayhew. Earliest (c. 1850) sociological study in English, describing myriad subcultures of London poor. Particularly remarkable for the thousands of pages of direct testimony taken from the lips of London prostitutes, thieves, beggars, street sellers, chimney-sweepers, street-musicians, "mudlarks," "pure-finders," rag-gatherers, "running-patterers," dock laborers, cab-men, and hundreds of others, quoted directly in this massive work. An extraordinarily vital picture of London emerges. 110 illustrations. Total of lxxvi + 1951pp. 6⅝ x 10.
21934-8, 21935-6, 21936-4, 21937-2 Four volumes, Paperbound $16.00

HISTORY OF THE LATER ROMAN EMPIRE, J. B. Bury. Eloquent, detailed reconstruction of Western and Byzantine Roman Empire by a major historian, from the death of Theodosius I (395 A.D.) to the death of Justinian (565). Extensive quotations from contemporary sources; full coverage of important Roman and foreign figures of the time. xxxiv + 965pp. 20398-0, 20399-9 Two volumes, Paperbound $7.00

AN INTELLECTUAL AND CULTURAL HISTORY OF THE WESTERN WORLD, Harry Elmer Barnes. Monumental study, tracing the development of the accomplishments that make up human culture. Every aspect of man's achievement surveyed from its origins in the Paleolithic to the present day (1964); social structures, ideas, economic systems, art, literature, technology, mathematics, the sciences, medicine, religion, jurisprudence, etc. Evaluations of the contributions of scores of great men. 1964 edition, revised and edited by scholars in the many fields represented. Total of xxix + 1381pp. 21275-0, 21276-9, 21277-7 Three volumes, Paperbound $10.50

LAST AND FIRST MEN AND STAR MAKER, TWO SCIENCE FICTION NOVELS, Olaf Stapledon. Greatest future histories in science fiction. In the first, human intelligence is the "hero," through strange paths of evolution, interplanetary invasions, incredible technologies, near extinctions and reemergences. Star Maker describes the quest of a band of star rovers for intelligence itself, through time and space: weird inhuman civilizations, crustacean minds, symbiotic worlds, etc. Complete, unabridged. v + 438pp.   (USO) 21962-3 Paperbound $3.00

THREE PROPHETIC NOVELS, H. G. Wells. Stages of a consistently planned future for mankind. *When the Sleeper Wakes,* and *A Story of the Days to Come,* anticipate *Brave New World* and *1984,* in the 21st Century; *The Time Machine,* only complete version in print, shows farther future and the end of mankind. All show Wells's greatest gifts as storyteller and novelist. Edited by E. F. Bleiler. x + 335pp.   (USO) 20605-X Paperbound $3.00

THE DEVIL'S DICTIONARY, Ambrose Bierce. America's own Oscar Wilde—Ambrose Bierce—offers his barbed iconoclastic wisdom in over 1,000 definitions hailed by H. L. Mencken as "some of the most gorgeous witticisms in the English language." 145pp.   20487-1 Paperbound $1.50

MAX AND MORITZ, Wilhelm Busch. Great children's classic, father of comic strip, of two bad boys, Max and Moritz. Also Ker and Plunk (Plisch und Plumm), Cat and Mouse, Deceitful Henry, Ice-Peter, The Boy and the Pipe, and five other pieces. Original German, with English translation. Edited by H. Arthur Klein; translations by various hands and H. Arthur Klein. vi + 216pp.
                          20181-3 Paperbound $2.00

PIGS IS PIGS AND OTHER FAVORITES, Ellis Parker Butler. The title story is one of the best humor short stories, as Mike Flannery obfuscates biology and English. Also included, That Pup of Murchison's, The Great American Pie Company, and Perkins of Portland. 14 illustrations. v + 109pp.   21532-6 Paperbound $1.50

THE PETERKIN PAPERS, Lucretia P. Hale. It takes genius to be as stupidly mad as the Peterkins, as they decide to become wise, celebrate the "Fourth," keep a cow, and otherwise strain the resources of the Lady from Philadelphia. Basic book of American humor. 153 illustrations. 219pp.   20794-3 Paperbound $2.00

PERRAULT'S FAIRY TALES, translated by A. E. Johnson and S. R. Littlewood, with 34 full-page illustrations by Gustave Doré. All the original Perrault stories—Cinderella, Sleeping Beauty, Bluebeard, Little Red Riding Hood, Puss in Boots, Tom Thumb, etc.—with their witty verse morals and the magnificent illustrations of Doré. One of the five or six great books of European fairy tales. viii + 117pp. 8⅛ x 11.   22311-6 Paperbound $2.00

OLD HUNGARIAN FAIRY TALES, Baroness Orczy. Favorites translated and adapted by author of the *Scarlet Pimpernel.* Eight fairy tales include "The Suitors of Princess Fire-Fly," "The Twin Hunchbacks," "Mr. Cuttlefish's Love Story," and "The Enchanted Cat." This little volume of magic and adventure will captivate children as it has for generations. 90 drawings by Montagu Barstow. 96pp.
                          (USO) 22293-4 Paperbound $1.95

THE PHILOSOPHY OF THE UPANISHADS, Paul Deussen. Clear, detailed statement of upanishadic system of thought, generally considered among best available. History of these works, full exposition of system emergent from them, parallel concepts in the West. Translated by A. S. Geden. xiv + 429pp.
21616-0 Paperbound $3.50

LANGUAGE, TRUTH AND LOGIC, Alfred J. Ayer. Famous, remarkably clear introduction to the Vienna and Cambridge schools of Logical Positivism; function of philosophy, elimination of metaphysical thought, nature of analysis, similar topics. "Wish I had written it myself," Bertrand Russell. 2nd, 1946 edition. 160pp.
20010-8 Paperbound $1.50

THE GUIDE FOR THE PERPLEXED, Moses Maimonides. Great classic of medieval Judaism, major attempt to reconcile revealed religion (Pentateuch, commentaries) and Aristotelian philosophy. Enormously important in all Western thought. Unabridged Friedländer translation. 50-page introduction. lix + 414pp.
(USO) 20351-4 Paperbound $4.50

OCCULT AND SUPERNATURAL PHENOMENA, D. H. Rawcliffe. Full, serious study of the most persistent delusions of mankind: crystal gazing, mediumistic trance, stigmata, lycanthropy, fire walking, dowsing, telepathy, ghosts, ESP, etc., and their relation to common forms of abnormal psychology. Formerly *Illusions and Delusions of the Supernatural and the Occult.* iii + 551pp. 20503-7 Paperbound $4.00

THE EGYPTIAN BOOK OF THE DEAD: THE PAPYRUS OF ANI, E. A. Wallis Budge. Full hieroglyphic text, interlinear transliteration of sounds, word for word translation, then smooth, connected translation; Theban recension. Basic work in Ancient Egyptian civilization; now even more significant than ever for historical importance, dilation of consciousness, etc. clvi + 377pp. 6½ x 9¼.
21866-X Paperbound $4.95

PSYCHOLOGY OF MUSIC, Carl E. Seashore. Basic, thorough survey of everything known about psychology of music up to 1940's; essential reading for psychologists, musicologists. Physical acoustics; auditory apparatus; relationship of physical sound to perceived sound; role of the mind in sorting, altering, suppressing, creating sound sensations; musical learning, testing for ability, absolute pitch, other topics. Records of Caruso, Menuhin analyzed. 88 figures. xix + 408pp.
21851-1 Paperbound $3.50

THE I CHING (THE BOOK OF CHANGES), translated by James Legge. Complete translated text plus appendices by Confucius, of perhaps the most penetrating divination book ever compiled. Indispensable to all study of early Oriental civilizations. 3 plates. xxiii + 448pp. 21062-6 Paperbound $3.50

THE UPANISHADS, translated by Max Müller. Twelve classical upanishads: Chandogya, Kena, Aitareya, Kaushitaki, Isa, Katha, Mundaka, Taittiriyaka, Brhadaranyaka, Svetasvatara, Prasna, Maitriyana. 160-page introduction, analysis by Prof. Müller. Total of 670pp. 20992-X, 20993-8 Two volumes, Paperbound $7.50

VISUAL ILLUSIONS: THEIR CAUSES, CHARACTERISTICS, AND APPLICATIONS, Matthew Luckiesh. Thorough description and discussion of optical illusion, geometric and perspective, particularly; size and shape distortions, illusions of color, of motion; natural illusions; use of illusion in art and magic, industry, etc. Most useful today with op art, also for classical art. Scores of · effects illustrated. Introduction by William H. Ittleson. 100 illustrations. xxi + 252pp.

21530-X Paperbound $2.00

A HANDBOOK OF ANATOMY FOR ART STUDENTS, Arthur Thomson. Thorough, virtually exhaustive coverage of skeletal structure, musculature, etc. Full text, supplemented by anatomical diagrams and drawings and by photographs of undraped figures. Unique in its comparison of male and female forms, pointing out differences of contour, texture, form. 211 figures, 40 drawings, 86 photographs. xx + 459pp. 5⅜ x 8⅜.

21163-0 Paperbound $3.50

150 MASTERPIECES OF DRAWING, Selected by Anthony Toney. Full page reproductions of drawings from the early 16th to the end of the 18th century, all beautifully reproduced: Rembrandt, Michelangelo, Dürer, Fragonard, Urs, Graf, Wouwerman, many others. First-rate browsing book, model book for artists. xviii + 150pp. 8⅜ x 11¼.

21032-4 Paperbound¹ $2.50

THE LATER WORK OF AUBREY BEARDSLEY, Aubrey Beardsley. Exotic, erotic, ironic masterpieces in full maturity: Comedy Ballet, Venus and Tannhauser, Pierrot, Lysistrata, Rape of the Lock, Savoy material, Ali Baba, Volpone, etc. This material revolutionized the art world, and is still powerful, fresh, brilliant. With *The Early Work,* all Beardsley's finest work. 174 plates, 2 in color. xiv + 176pp. 8⅛ x 11.

21817-1 Paperbound $3.75

DRAWINGS OF REMBRANDT, Rembrandt van Rijn. Complete reproduction of fabulously rare edition by Lippmann and Hofstede de Groot, completely reedited, updated, improved by Prof. Seymour Slive, Fogg Museum. Portraits, Biblical sketches, landscapes, Oriental types, nudes, episodes from classical mythology—All Rembrandt's fertile genius. Also selection of drawings by his pupils and followers. "Stunning volumes," *Saturday Review.* 550 illustrations. lxxviii + 552pp. 9⅛ x 12¼.

21485-0, 21486-9 Two volumes, Paperbound $10.00

THE DISASTERS OF WAR, Francisco Goya. One of the masterpieces of Western civilization—83 etchings that record Goya's shattering, bitter reaction to the Napoleonic war that swept through Spain after the insurrection of 1808 and to war in general. Reprint of the first edition, with three additional plates from Boston's Museum of Fine Arts. All plates facsimile size. Introduction by Philip Hofer, Fogg Museum. v + 97pp. 9⅜ x 8¼.

21872-4 Paperbound $2.50

GRAPHIC WORKS OF ODILON REDON. Largest collection of Redon's graphic works ever assembled: 172 lithographs, 28 etchings and engravings, 9 drawings. These include some of his most famous works. All the plates from *Odilon Redon: oeuvre graphique complet,* plus additional plates. New introduction and caption translations by Alfred Werner. 209 illustrations. xxvii + 209pp. 9⅛ x 12¼.

21966-8 Paperbound $4.50

INCIDENTS OF TRAVEL IN YUCATAN, John L. Stephens. Classic (1843) exploration of jungles of Yucatan, looking for evidences of Maya civilization. Stephens found many ruins; comments on travel adventures, Mexican and Indian culture. 127 striking illustrations by F. Catherwood. Total of 669 pp.

20926-1, 20927-X Two volumes, Paperbound $5.50

INCIDENTS OF TRAVEL IN CENTRAL AMERICA, CHIAPAS, AND YUCATAN, John L. Stephens. An exciting travel journal and an important classic of archeology. Narrative relates his almost single-handed discovery of the Mayan culture, and exploration of the ruined cities of Copan, Palenque, Utatlan and others; the monuments they dug from the earth, the temples buried in the jungle, the customs of poverty-stricken Indians living a stone's throw from the ruined palaces. 115 drawings by F. Catherwood. Portrait of Stephens. xii + 812pp.

22404-X, 22405-8 Two volumes, Paperbound $6.00

A NEW VOYAGE ROUND THE WORLD, William Dampier. Late 17-century naturalist joined the pirates of the Spanish Main to gather information; remarkably vivid account of buccaneers, pirates; detailed, accurate account of botany, zoology, ethnography of lands visited. Probably the most important early English voyage, enormous implications for British exploration, trade, colonial policy. Also most interesting reading. Argonaut edition, introduction by Sir Albert Gray. New introduction by Percy Adams. 6 plates, 7 illustrations. xlvii + 376pp. 6½ x 9¼.

21900-3 Paperbound $3.00

INTERNATIONAL AIRLINE PHRASE BOOK IN SIX LANGUAGES, Joseph W. Bátor. Important phrases and sentences in English paralleled with French, German, Portuguese, Italian, Spanish equivalents, covering all possible airport-travel situations; created for airline personnel as well as tourist by Language Chief, Pan American Airlines. xiv + 204pp.

22017-6 Paperbound $2.25

STAGE COACH AND TAVERN DAYS, Alice Morse Earle. Detailed, lively account of the early days of taverns; their uses and importance in the social, political and military life; furnishings and decorations; locations; food and drink; tavern signs, etc. Second half covers every aspect of early travel; the roads, coaches, drivers, etc. Nostalgic, charming, packed with fascinating material. 157 illustrations, mostly photographs. xiv + 449pp.

22518-6 Paperbound $4.00

NORSE DISCOVERIES AND EXPLORATIONS IN NORTH AMERICA, Hjalmar R. Holand. The perplexing Kensington Stone, found in Minnesota at the end of the 19th century. Is it a record of a Scandinavian expedition to North America in the 14th century? Or is it one of the most successful hoaxes in history. A scientific detective investigation. Formerly *Westward from Vinland*. 31 photographs, 17 figures. x + 354pp.

22014-1 Paperbound $2.75

A BOOK OF OLD MAPS, compiled and edited by Emerson D. Fite and Archibald Freeman. 74 old maps offer an unusual survey of the discovery, settlement and growth of America down to the close of the Revolutionary war: maps showing Norse settlements in Greenland, the explorations of Columbus, Verrazano, Cabot, Champlain, Joliet, Drake, Hudson, etc., campaigns of Revolutionary war battles, and much more. Each map is accompanied by a brief historical essay. xvi + 299pp. 11 x 13¾.

22084-2 Paperbound $7.00

"ESSENTIAL GRAMMAR" SERIES

All you really need to know about modern, colloquial grammar. Many educational shortcuts help you learn faster, understand better. Detailed cognate lists teach you to recognize similarities between English and foreign words and roots—make learning vocabulary easy and interesting. Excellent for independent study or as a supplement to record courses.

ESSENTIAL FRENCH GRAMMAR, Seymour Resnick. 2500-item cognate list. 159pp.
(EBE) 20419-7 Paperbound $1.50

ESSENTIAL GERMAN GRAMMAR, Guy Stern and Everett F. Bleiler. Unusual shortcuts on noun declension, word order, compound verbs. 124pp.
(EBE) 20422-7 Paperbound $1.25

ESSENTIAL ITALIAN GRAMMAR, Olga Ragusa. 111pp.
(EBE) 20779-X Paperbound $1.25

ESSENTIAL JAPANESE GRAMMAR, Everett F. Bleiler. In Romaji transcription; no characters needed. Japanese grammar is regular and simple. 156pp.
21027-8 Paperbound $1.50

ESSENTIAL PORTUGUESE GRAMMAR, Alexander da R. Prista. vi + 114pp.
21650-0 Paperbound $1.35

ESSENTIAL SPANISH GRAMMAR, Seymour Resnick. 2500 word cognate list. 115pp.
(EBE) 20780-3 Paperbound $1.25

ESSENTIAL ENGLISH GRAMMAR, Philip Gucker. Combines best features of modern, functional and traditional approaches. For refresher, class use, home study. x + 177pp.
21649-7 Paperbound $1.75

A PHRASE AND SENTENCE DICTIONARY OF SPOKEN SPANISH. Prepared for U. S. War Department by U. S. linguists. As above, unit is idiom, phrase or sentence rather than word. English-Spanish and Spanish-English sections contain modern equivalents of over 18,000 sentences. Introduction and appendix as above. iv + 513pp.
20495-2 Paperbound $3.50

A PHRASE AND SENTENCE DICTIONARY OF SPOKEN RUSSIAN. Dictionary prepared for U. S. War Department by U. S. linguists. Basic unit is not the word, but the idiom, phrase or sentence. English-Russian and Russian-English sections contain modern equivalents for over 30,000 phrases. Grammatical introduction covers phonetics, writing, syntax. Appendix of word lists for food, numbers, geographical names, etc. vi + 573 pp. 6⅛ x 9¼.
20496-0 Paperbound $5.50

CONVERSATIONAL CHINESE FOR BEGINNERS, Morris Swadesh. Phonetic system, beginner's course in Pai Hua Mandarin Chinese covering most important, most useful speech patterns. Emphasis on modern colloquial usage. Formerly *Chinese in Your Pocket*. xvi + 158pp.
21123-1 Paperbound $1.75

JOHANN SEBASTIAN BACH, Philipp Spitta. One of the great classics of musicology, this definitive analysis of Bach's music (and life) has never been surpassed. Lucid, nontechnical analyses of hundreds of pieces (30 pages devoted to St. Matthew Passion, 26 to B Minor Mass). Also includes major analysis of 18th-century music. 450 musical examples. 40-page musical supplement. Total of xx + 1799pp.
(EUK) 22278-0, 22279-9 Two volumes, Clothbound $25.00

MOZART AND HIS PIANO CONCERTOS, Cuthbert Girdlestone. The only full-length study of an important area of Mozart's creativity. Provides detailed analyses of all 23 concertos, traces inspirational sources. 417 musical examples. Second edition. 509pp.
21271-8 Paperbound $4.50

THE PERFECT WAGNERITE: A COMMENTARY ON THE NIBLUNG'S RING, George Bernard Shaw. Brilliant and still relevant criticism in remarkable essays on Wagner's Ring cycle, Shaw's ideas on political and social ideology behind the plots, role of Leitmotifs, vocal requisites, etc. Prefaces. xxi + 136pp.
(USO) 21707-8 Paperbound $1.75

DON GIOVANNI, W. A. Mozart. Complete libretto, modern English translation; biographies of composer and librettist; accounts of early performances and critical reaction. Lavishly illustrated. All the material you need to understand and appreciate this great work. Dover Opera Guide and Libretto Series; translated and introduced by Ellen Bleiler. 92 illustrations. 209pp.
21134-7 Paperbound $2.00

BASIC ELECTRICITY, U. S. Bureau of Naval Personel. Originally a training course, best non-technical coverage of basic theory of electricity and its applications. Fundamental concepts, batteries, circuits, conductors and wiring techniques, AC and DC, inductance and capacitance, generators, motors, transformers, magnetic amplifiers, synchros, servomechanisms, etc. Also covers blue-prints, electrical diagrams, etc. Many questions, with answers. 349 illustrations. x + 448pp. 6½ x 9¼.
20973-3 Paperbound $3.50

REPRODUCTION OF SOUND, Edgar Villchur. Thorough coverage for laymen of high fidelity systems, reproducing systems in general, needles, amplifiers, preamps, loudspeakers, feedback, explaining physical background. "A rare talent for making technicalities vividly comprehensible," R. Darrell, *High Fidelity*. 69 figures. iv + 92pp.
21515-6 Paperbound $1.35

HEAR ME TALKIN' TO YA: THE STORY OF JAZZ AS TOLD BY THE MEN WHO MADE IT, Nat Shapiro and Nat Hentoff. Louis Armstrong, Fats Waller, Jo Jones, Clarence Williams, Billy Holiday, Duke Ellington, Jelly Roll Morton and dozens of other jazz greats tell how it was in Chicago's South Side, New Orleans, depression Harlem and the modern West Coast as jazz was born and grew. xvi + 429pp.
21726-4 Paperbound $3.95

FABLES OF AESOP, translated by Sir Roger L'Estrange. A reproduction of the very rare 1931 Paris edition; a selection of the most interesting fables, together with 50 imaginative drawings by Alexander Calder. v + 128pp. 6½x9¼.
21780-9 Paperbound $1.50

PLANETS, STARS AND GALAXIES: DESCRIPTIVE ASTRONOMY FOR BEGINNERS, A. E. Fanning. Comprehensive introductory survey of astronomy: the sun, solar system, stars, galaxies, universe, cosmology; up-to-date, including quasars, radio stars, etc. Preface by Prof. Donald Menzel. 24pp. of photographs. 189pp. 5¼ x 8¼.
21680-2 Paperbound $2.50

TEACH YOURSELF CALCULUS, P. Abbott. With a good background in algebra and trig, you can teach yourself calculus with this book. Simple, straightforward introduction to functions of all kinds, integration, differentiation, series, etc. "Students who are beginning to study calculus method will derive great help from this book." Faraday House Journal. 308pp.
20683-1 Clothbound $2.50

TEACH YOURSELF TRIGONOMETRY, P. Abbott. Geometrical foundations, indices and logarithms, ratios, angles, circular measure, etc. are presented in this sound, easy-to-use text. Excellent for the beginner or as a brush up, this text carries the student through the solution of triangles. 204pp.
20682-3 Clothbound $2.00

BASIC MACHINES AND HOW THEY WORK, U. S. Bureau of Naval Personnel. Originally used in U.S. Naval training schools, this book clearly explains the operation of a progression of machines, from the simplest—lever, wheel and axle, inclined plane, wedge, screw—to the most complex—typewriter, internal combustion engine, computer mechanism. Utilizing an approach that requires only an elementary understanding of mathematics, these explanations build logically upon each other and are assisted by over 200 drawings and diagrams. Perfect as a technical school manual or as a self-teaching aid to the layman. 204 figures. Preface. Index. vii + 161pp. 6½ x 9¼.
21709-4 Paperbound $2.50

THE FRIENDLY STARS, Martha Evans Martin. Classic has taught naked-eye observation of stars, planets to hundreds of thousands, still not surpassed for charm, lucidity, adequacy. Completely updated by Professor Donald H. Menzel, Harvard Observatory. 25 illustrations. 16 x 30 chart. x + 147pp.
21099-5 Paperbound $2.00

MUSIC OF THE SPHERES: THE MATERIAL UNIVERSE FROM ATOM TO QUASAR, SIMPLY EXPLAINED, Guy Murchie. Extremely broad, brilliantly written popular account begins with the solar system and reaches to dividing line between matter and nonmatter; latest understandings presented with exceptional clarity. Volume One: Planets, stars, galaxies, cosmology, geology, celestial mechanics, latest astronomical discoveries; Volume Two: Matter, atoms, waves, radiation, relativity, chemical action, heat, nuclear energy, quantum theory, music, light, color, probability, antimatter, antigravity, and similar topics. 319 figures. 1967 (second) edition. Total of xx + 644pp.
21809-0, 21810-4 Two volumes, Paperbound $5.75

OLD-TIME SCHOOLS AND SCHOOL BOOKS, Clifton Johnson. Illustrations and rhymes from early primers, abundant quotations from early textbooks, many anecdotes of school life enliven this study of elementary schools from Puritans to middle 19th century. Introduction by Carl Withers. 234 illustrations. xxxiii + 381pp.
21031-6 Paperbound $4.00

EAST O' THE SUN AND WEST O' THE MOON, George W. Dasent. Considered the best of all translations of these Norwegian folk tales, this collection has been enjoyed by generations of children (and folklorists too). Includes True and Untrue, Why the Sea is Salt, East O' the Sun and West O' the Moon, Why the Bear is Stumpy-Tailed, Boots and the Troll, The Cock and the Hen, Rich Peter the Pedlar, and 52 more. The only edition with all 59 tales. 77 illustrations by Erik Werenskiold and Theodor Kittelsen. xv + 418pp. 22521-6 Paperbound $3.50

GOOPS AND HOW TO BE THEM, Gelett Burgess. Classic of tongue-in-cheek humor, masquerading as etiquette book. 87 verses, twice as many cartoons, show mischievous Goops as they demonstrate to children virtues of table manners, neatness, courtesy, etc. Favorite for generations. viii + 88pp. 6½ x 9¼.
22233-0 Paperbound $1.50

ALICE'S ADVENTURES UNDER GROUND, Lewis Carroll. The first version, quite different from the final Alice in Wonderland, printed out by Carroll himself with his own illustrations. Complete facsimile of the "million dollar" manuscript Carroll gave to Alice Liddell in 1864. Introduction by Martin Gardner. viii + 96pp. Title and dedication pages in color. 21482-6 Paperbound $1.25

THE BROWNIES, THEIR BOOK, Palmer Cox. Small as mice, cunning as foxes, exuberant and full of mischief, the Brownies go to the zoo, toy shop, seashore, circus, etc., in 24 verse adventures and 266 illustrations. Long a favorite, since their first appearance in St. Nicholas Magazine. xi + 144pp. 6⅝ x 9¼.
21265-3 Paperbound $1.75

SONGS OF CHILDHOOD, Walter De La Mare. Published (under the pseudonym Walter Ramal) when De La Mare was only 29, this charming collection has long been a favorite children's book. A facsimile of the first edition in paper, the 47 poems capture the simplicity of the nursery rhyme and the ballad, including such lyrics as I Met Eve, Tartary, The Silver Penny. vii + 106pp. (USO) 21972-0 Paperbound $1.25

THE COMPLETE NONSENSE OF EDWARD LEAR, Edward Lear. The finest 19th-century humorist-cartoonist in full: all nonsense limericks, zany alphabets, Owl and Pussycat, songs, nonsense botany, and more than 500 illustrations by Lear himself. Edited by Holbrook Jackson. xxix + 287pp. (USO) 20167-8 Paperbound $2.00

BILLY WHISKERS: THE AUTOBIOGRAPHY OF A GOAT, Frances Trego Montgomery. A favorite of children since the early 20th century, here are the escapades of that rambunctious, irresistible and mischievous goat—Billy Whiskers. Much in the spirit of Peck's Bad Boy, this is a book that children never tire of reading or hearing. All the original familiar illustrations by W. H. Fry are included: 6 color plates, 18 black and white drawings. 159pp. 22345-0 Paperbound $2.00

MOTHER GOOSE MELODIES. Faithful republication of the fabulously rare Munroe and Francis "copyright 1833" Boston edition—the most important Mother Goose collection, usually referred to as the "original." Familiar rhymes plus many rare ones, with wonderful old woodcut illustrations. Edited by E. F. Bleiler. 128pp. 4½ x 6⅜. 22577-1 Paperbound $1.00

AGAINST THE GRAIN (A REBOURS), Joris K. Huysmans. Filled with weird images, evidences of a bizarre imagination, exotic experiments with hallucinatory drugs, rich tastes and smells and the diversions of its sybarite hero Duc Jean des Esseintes, this classic novel pushed 19th-century literary decadence to its limits. Full unabridged edition. Do not confuse this with abridged editions generally sold. Introduction by Havelock Ellis. xlix + 206pp. 22190-3 Paperbound $2.50

VARIORUM SHAKESPEARE: HAMLET. Edited by Horace H. Furness; a landmark of American scholarship. Exhaustive footnotes and appendices treat all doubtful words and phrases, as well as suggested critical emendations throughout the play's history. First volume contains editor's own text, collated with all Quartos and Folios. Second volume contains full first Quarto, translations of Shakespeare's sources (Belleforest, and Saxo Grammaticus), Der Bestrafte Brudermord, and many essays on critical and historical points of interest by major authorities of past and present. Includes details of staging and costuming over the years. By far the best edition available for serious students of Shakespeare. Total of xx + 905pp. 21004-9, 21005-7, 2 volumes, Paperbound $7.00

A LIFE OF WILLIAM SHAKESPEARE, Sir Sidney Lee. This is the standard life of Shakespeare, summarizing everything known about Shakespeare and his plays. Incredibly rich in material, broad in coverage, clear and judicious, it has served thousands as the best introduction to Shakespeare. 1931 edition. 9 plates. xxix + 792pp. 21967-4 Paperbound $4.50

MASTERS OF THE DRAMA, John Gassner. Most comprehensive history of the drama in print, covering every tradition from Greeks to modern Europe and America, including India, Far East, etc. Covers more than 800 dramatists, 2000 plays, with biographical material, plot summaries, theatre history, criticism, etc. "Best of its kind in English," *New Republic*. 77 illustrations. xxii + 890pp. 20100-7 Clothbound $10.00

THE EVOLUTION OF THE ENGLISH LANGUAGE, George McKnight. The growth of English, from the 14th century to the present. Unusual, non-technical account presents basic information in very interesting form: sound shifts, change in grammar and syntax, vocabulary growth, similar topics. Abundantly illustrated with quotations. Formerly *Modern English in the Making*. xii + 590pp. 21932-1 Paperbound $3.50

AN ETYMOLOGICAL DICTIONARY OF MODERN ENGLISH, Ernest Weekley. Fullest, richest work of its sort, by foremost British lexicographer. Detailed word histories, including many colloquial and archaic words; extensive quotations. Do not confuse this with the Concise Etymological Dictionary, which is much abridged. Total of xxvii + 830pp. 6½ x 9¼. 21873-2, 21874-0 Two volumes, Paperbound $7.90

FLATLAND: A ROMANCE OF MANY DIMENSIONS, E. A. Abbott. Classic of science-fiction explores ramifications of life in a two-dimensional world, and what happens when a three-dimensional being intrudes. Amusing reading, but also useful as introduction to thought about hyperspace. Introduction by Banesh Hoffmann. 16 illustrations. xx + 103pp. 20001-9 Paperbound $1.00

POEMS OF ANNE BRADSTREET, edited with an introduction by Robert Hutchinson. A new selection of poems by America's first poet and perhaps the first significant woman poet in the English language. 48 poems display her development in works of considerable variety—love poems, domestic poems, religious meditations, formal elegies, "quaternions," etc. Notes, bibliography. viii + 222pp.

22160-1 Paperbound $2.50

THREE GOTHIC NOVELS: THE CASTLE OF OTRANTO BY HORACE WALPOLE; VATHEK BY WILLIAM BECKFORD; THE VAMPYRE BY JOHN POLIDORI, WITH FRAGMENT OF A NOVEL BY LORD BYRON, edited by E. F. Bleiler. The first Gothic novel, by Walpole; the finest Oriental tale in English, by Beckford; powerful Romantic supernatural story in versions by Polidori and Byron. All extremely important in history of literature; all still exciting, packed with supernatural thrills, ghosts, haunted castles, magic, etc. xl + 291pp.

21232-7 Paperbound $3.00

THE BEST TALES OF HOFFMANN, E. T. A. Hoffmann. 10 of Hoffmann's most important stories, in modern re-editings of standard translations: Nutcracker and the King of Mice, Signor Formica, Automata, The Sandman, Rath Krespel, The Golden Flowerpot, Master Martin the Cooper, The Mines of Falun, The King's Betrothed, A New Year's Eve Adventure. 7 illustrations by Hoffmann. Edited by E. F. Bleiler. xxxix + 419pp.   21793-0 Paperbound $3.00

GHOST AND HORROR STORIES OF AMBROSE BIERCE, Ambrose Bierce. 23 strikingly modern stories of the horrors latent in the human mind: The Eyes of the Panther, The Damned Thing, An Occurrence at Owl Creek Bridge, An Inhabitant of Carcosa, etc., plus the dream-essay, Visions of the Night. Edited by E. F. Bleiler. xxii + 199pp.   20767-6 Paperbound $2.00

BEST GHOST STORIES OF J. S. LEFANU, J. Sheridan LeFanu. Finest stories by Victorian master often considered greatest supernatural writer of all. Carmilla, Green Tea, The Haunted Baronet, The Familiar, and 12 others. Most never before available in the U. S. A. Edited by E. F. Bleiler. 8 illustrations from Victorian publications. xvii + 467pp.   20415-4 Paperbound $3.00

MATHEMATICAL FOUNDATIONS OF INFORMATION THEORY, A. I. Khinchin. Comprehensive introduction to work of Shannon, McMillan, Feinstein and Khinchin, placing these investigations on a rigorous mathematical basis. Covers entropy concept in probability theory, uniqueness theorem, Shannon's inequality, ergodic sources, the E property, martingale concept, noise, Feinstein's fundamental lemma, Shanon's first and second theorems. Translated by R. A. Silverman and M. D. Friedman. iii + 120pp.   60434-9 Paperbound $2.00

SEVEN SCIENCE FICTION NOVELS, H. G. Wells. The standard collection of the great novels. Complete, unabridged. *First Men in the Moon, Island of Dr. Moreau, War of the Worlds, Food of the Gods, Invisible Man, Time Machine, In the Days of the Comet.* Not only science fiction fans, but every educated person owes it to himself to read these novels. 1015pp.   (USO) 20264-X Clothbound $6.00

TWO LITTLE SAVAGES; BEING THE ADVENTURES OF TWO BOYS WHO LIVED AS INDIANS AND WHAT THEY LEARNED, Ernest Thompson Seton. Great classic of nature and boyhood provides a vast range of woodlore in most palatable form, a genuinely entertaining story. Two farm boys build a teepee in woods and live in it for a month, working out Indian solutions to living problems, star lore, birds and animals, plants, etc. 293 illustrations. vii + 286pp.

20985-7 Paperbound $2.50

PETER PIPER'S PRACTICAL PRINCIPLES OF PLAIN & PERFECT PRONUNCIATION. Alliterative jingles and tongue-twisters of surprising charm, that made their first appearance in America about 1830. Republished in full with the spirited woodcut illustrations from this earliest American edition. 32pp. 4½ x 6⅜.

22560-7 Paperbound $1.00

SCIENCE EXPERIMENTS AND AMUSEMENTS FOR CHILDREN, Charles Vivian. 73 easy experiments, requiring only materials found at home or easily available, such as candles, coins, steel wool, etc.; illustrate basic phenomena like vacuum, simple chemical reaction, etc. All safe. Modern, well-planned. Formerly *Science Games for Children*. 102 photos, numerous drawings. 96pp. 6⅛ x 9¼.

21856-2 Paperbound $1.25

AN INTRODUCTION TO CHESS MOVES AND TACTICS SIMPLY EXPLAINED, Leonard Barden. Informal intermediate introduction, quite strong in explaining reasons for moves. Covers basic material, tactics, important openings, traps, positional play in middle game, end game. Attempts to isolate patterns and recurrent configurations. Formerly *Chess*. 58 figures. 102pp. (USO) 21210-6 Paperbound $1.25

LASKER'S MANUAL OF CHESS, Dr. Emanuel Lasker. Lasker was not only one of the five great World Champions, he was also one of the ablest expositors, theorists, and analysts. In many ways, his Manual, permeated with his philosophy of battle, filled with keen insights, is one of the greatest works ever written on chess. Filled with analyzed games by the great players. A single-volume library that will profit almost any chess player, beginner or master. 308 diagrams. xli x 349pp.

20640-8 Paperbound $2.75

THE MASTER BOOK OF MATHEMATICAL RECREATIONS, Fred Schuh. In opinion of many the finest work ever prepared on mathematical puzzles, stunts, recreations; exhaustively thorough explanations of mathematics involved, analysis of effects, citation of puzzles and games. Mathematics involved is elementary. Translated bv F. Göbel. 194 figures. xxiv + 430pp. 22134-2 Paperbound $4.00

MATHEMATICS, MAGIC AND MYSTERY, Martin Gardner. Puzzle editor for Scientific American explains mathematics behind various mystifying tricks: card tricks, stage "mind reading," coin and match tricks, counting out games, geometric dissections, etc. Probability sets, theory of numbers clearly explained. Also provides more than 400 tricks, guaranteed to work, that you can do. 135 illustrations. xii + 176pp.

20335-2 Paperbound $2.00

THE ARCHITECTURE OF COUNTRY HOUSES, Andrew J. Downing. Together with Vaux's *Villas and Cottages* this is the basic book for Hudson River Gothic architecture of the middle Victorian period. Full, sound discussions of general aspects of housing, architecture, style, decoration, furnishing, together with scores of detailed house plans, illustrations of specific buildings, accompanied by full text. Perhaps the most influential single American architectural book. 1850 edition. Introduction by J. Stewart Johnson. 321 figures, 34 architectural designs. xvi + 560pp.
22003-6 Paperbound $5.00

LOST EXAMPLES OF COLONIAL ARCHITECTURE, John Mead Howells. Full-page photographs of buildings that have disappeared or been so altered as to be denatured, including many designed by major early American architects. 245 plates. xvii + 248pp. 7⅞ x 10¾.
21143-6 Paperbound $3.50

DOMESTIC ARCHITECTURE OF THE AMERICAN COLONIES AND OF THE EARLY REPUBLIC, Fiske Kimball. Foremost architect and restorer of Williamsburg and Monticello covers nearly 200 homes between 1620-1825. Architectural details, construction, style features, special fixtures, floor plans, etc. Generally considered finest work in its area. 219 illustrations of houses, doorways, windows, capital mantels. xx + 314pp. 7⅞ x 10¾.
21743-4 Paperbound $4.00

EARLY AMERICAN ROOMS: 1650-1858, edited by Russell Hawes Kettell. Tour of 12 rooms, each representative of a different era in American history and each furnished, decorated, designed and occupied in the style of the era. 72 plans and elevations, 8-page color section, etc., show fabrics, wall papers, arrangements, etc. Full descriptive text. xvii + 200pp. of text. 8⅜ x 11¼.
21633-0 Paperbound $5.00

THE FITZWILLIAM VIRGINAL BOOK, edited by J. Fuller Maitland and W. B. Squire. Full modern printing of famous early 17th-century ms. volume of 300 works by Morley, Byrd, Bull, Gibbons, etc. For piano or other modern keyboard instrument; easy to read format. xxxvi + 938pp. 8⅜ x 11.
21068-5, 21069-3 Two volumes, Paperbound $12.00

KEYBOARD MUSIC, Johann Sebastian Bach. Bach Gesellschaft edition. A rich selection of Bach's masterpieces for the harpsichord: the six English Suites, six French Suites, the six Partitas (Clavierübung part I), the Goldberg Variations (Clavierübung part IV), the fifteen Two-Part Inventions and the fifteen Three-Part Sinfonias. Clearly reproduced on large sheets with ample margins; eminently playable. vi + 312pp. 8⅛ x 11.
22360-4 Paperbound $5.00

THE MUSIC OF BACH: AN INTRODUCTION, Charles Sanford Terry. A fine, nontechnical introduction to Bach's music, both instrumental and vocal. Covers organ music, chamber music, passion music, other types. Analyzes themes, developments, innovations. x + 114pp.
21075-8 Paperbound $1.95

BEETHOVEN AND HIS NINE SYMPHONIES, Sir George Grove. Noted British musicologist provides best history, analysis, commentary on symphonies. Very thorough, rigorously accurate; necessary to both advanced student and amateur music lover. 436 musical passages. vii + 407 pp.
20334-4 Paperbound $4.00

THE PRINCIPLES OF PSYCHOLOGY, William James. The famous long course, complete and unabridged. Stream of thought, time perception, memory, experimental methods—these are only some of the concerns of a work that was years ahead of its time and still valid, interesting, useful. 94 figures. Total of xviii + 1391pp.
20381-6, 20382-4 Two volumes, Paperbound $9.00

THE STRANGE STORY OF THE QUANTUM, Banesh Hoffmann. Non-mathematical but thorough explanation of work of Planck, Einstein, Bohr, Pauli, de Broglie, Schrödinger, Heisenberg, Dirac, Feynman, etc. No technical background needed. "Of books attempting such an account, this is the best," Henry Margenau, Yale. 40-page "Postscript 1959." xii + 285pp.
20518-5 Paperbound $3.00

THE RISE OF THE NEW PHYSICS, A. d'Abro. Most thorough explanation in print of central core of mathematical physics, both classical and modern; from Newton to Dirac and Heisenberg. Both history and exposition; philosophy of science, causality, explanations of higher mathematics, analytical mechanics, electromagnetism, thermo-dynamics, phase rule, special and general relativity, matrices. No higher mathematics needed to follow exposition, though treatment is elementary to intermediate in level. Recommended to serious student who wishes verbal understanding. 97 illustrations. xvii + 982pp.
20003-5, 20004-3 Two volumes, Paperbound $10.00

GREAT IDEAS OF OPERATIONS RESEARCH, Jagjit Singh. Easily followed non-technical explanation of mathematical tools, aims, results: statistics, linear programming, game theory, queueing theory, Monte Carlo simulation, etc. Uses only elementary mathematics. Many case studies, several analyzed in detail. Clarity, breadth make this excellent for specialist in another field who wishes background. 41 figures. x + 228pp.
21886-4 Paperbound $2.50

GREAT IDEAS OF MODERN MATHEMATICS: THEIR NATURE AND USE, Jagjit Singh. Internationally famous expositor, winner of Unesco's Kalinga Award for science popularization explains verbally such topics as differential equations, matrices, groups, sets, transformations, mathematical logic and other important modern mathematics, as well as use in physics, astrophysics, and similar fields. Superb exposition for layman, scientist in other areas. viii + 312pp.
20587-8 Paperbound $2.75

GREAT IDEAS IN INFORMATION THEORY, LANGUAGE AND CYBERNETICS, Jagjit Singh. The analog and digital computers, how they work, how they are like and unlike the human brain, the men who developed them, their future applications, computer terminology. An essential book for today, even for readers with little math. Some mathematical demonstrations included for more advanced readers. 118 figures. Tables. ix + 338pp.
21694-2 Paperbound $2.50

CHANCE, LUCK AND STATISTICS, Horace C. Levinson. Non-mathematical presentation of fundamentals of probability theory and science of statistics and their applications. Games of chance, betting odds, misuse of statistics, normal and skew distributions, birth rates, stock speculation, insurance. Enlarged edition. Formerly "The Science of Chance." xiii + 357pp.
21007-3 Paperbound $2.50

THE RED FAIRY BOOK, Andrew Lang. Lang's color fairy books have long been children's favorites. This volume includes Rapunzel, Jack and the Bean-stalk and 35 other stories, familiar and unfamiliar. 4 plates, 93 illustrations x + 367pp.
21673-X Paperbound $2.50

THE BLUE FAIRY BOOK, Andrew Lang. Lang's tales come from all countries and all times. Here are 37 tales from Grimm, the Arabian Nights, Greek Mythology, and other fascinating sources. 8 plates, 130 illustrations. xi + 390pp.
21437-0 Paperbound $2.75

HOUSEHOLD STORIES BY THE BROTHERS GRIMM. Classic English-language edition of the well-known tales — Rumpelstiltskin, Snow White, Hansel and Gretel, The Twelve Brothers, Faithful John, Rapunzel, Tom Thumb (52 stories in all). Translated into simple, straightforward English by Lucy Crane. Ornamented with headpieces, vignettes, elaborate decorative initials and a dozen full-page illustrations by Walter Crane. x + 269pp.
21080-4 Paperbound **$2.00**

THE MERRY ADVENTURES OF ROBIN HOOD, Howard Pyle. The finest modern versions of the traditional ballads and tales about the great English outlaw. Howard Pyle's complete prose version, with every word, every illustration of the first edition. Do not confuse this facsimile of the original (1883) with modern editions that change text or illustrations. 23 plates plus many page decorations. xxii + 296pp.
22043-5 Paperbound $2.75

THE STORY OF KING ARTHUR AND HIS KNIGHTS, Howard Pyle. The finest children's version of the life of King Arthur; brilliantly retold by Pyle, with 48 of his most imaginative illustrations. xviii + 313pp. 6⅛ x 9¼.
21445-1 Paperbound $2.50

THE WONDERFUL WIZARD OF OZ, L. Frank Baum. America's finest children's book in facsimile of first edition with all Denslow illustrations in full color. The edition a child should have. Introduction by Martin Gardner. 23 color plates, scores of drawings. iv + 267pp.
20691-2 Paperbound $3.50

THE MARVELOUS LAND OF OZ, L. Frank Baum. The second Oz book, every bit as imaginative as the Wizard. The hero is a boy named Tip, but the Scarecrow and the Tin Woodman are back, as is the Oz magic. 16 color plates, 120 drawings by John R. Neill. 287pp.
20692-0 Paperbound $2.50

THE MAGICAL MONARCH OF MO, L. Frank Baum. Remarkable adventures in a land even stranger than Oz. The best of Baum's books not in the Oz series. 15 color plates and dozens of drawings by Frank Verbeck. xviii + 237pp.
21892-9 Paperbound $2.25

THE BAD CHILD'S BOOK OF BEASTS, MORE BEASTS FOR WORSE CHILDREN, A MORAL ALPHABET, Hilaire Belloc. Three complete humor classics in one volume. Be kind to the frog, and do not call him names . . . and 28 other whimsical animals. Familiar favorites and some not so well known. Illustrated by Basil Blackwell. 156pp.
(USO) 20749-8 Paperbound $1.50

AMERICAN FOOD AND GAME FISHES, David S. Jordan and Barton W. Evermann. Definitive source of information, detailed and accurate enough to enable the sportsman and nature lover to identify conclusively some 1,000 species and sub-species of North American fish, sought for food or sport. Coverage of range, physiology, habits, life history, food value. Best methods of capture, interest to the angler, advice on bait, fly-fishing, etc. 338 drawings and photographs. l + 574pp. 6⅝ x 9⅜.

22196-2 Paperbound $5.00

THE FROG BOOK, Mary C. Dickerson. Complete with extensive finding keys, over 300 photographs, and an introduction to the general biology of frogs and toads, this is the classic non-technical study of Northeastern and Central species. 58 species; 290 photographs and 16 color plates. xvii + 253pp.

21973-9 Paperbound $4.00

THE MOTH BOOK: A GUIDE TO THE MOTHS OF NORTH AMERICA, William J. Holland. Classical study, eagerly sought after and used for the past 60 years. Clear identification manual to more than 2,000 different moths, largest manual in existence. General information about moths, capturing, mounting, classifying, etc., followed by species by species descriptions. 263 illustrations plus 48 color plates show almost every species, full size. 1968 edition, preface, nomenclature changes by A. E. Brower. xxiv + 479pp. of text. 6½ x 9¼.

21948-8 Paperbound $6.00

THE SEA-BEACH AT EBB-TIDE, Augusta Foote Arnold. Interested amateur can identify hundreds of marine plants and animals on coasts of North America; marine algae; seaweeds; squids; hermit crabs; horse shoe crabs; shrimps; corals; sea anemones; etc. Species descriptions cover: structure; food; reproductive cycle; size; shape; color; habitat; etc. Over 600 drawings. 85 plates. xii + 490pp.

21949-6 Paperbound $4.00

COMMON BIRD SONGS, Donald J. Borror. 33⅓ 12-inch record presents songs of 60 important birds of the eastern United States. A thorough, serious record which provides several examples for each bird, showing different types of song, individual variations, etc. Inestimable identification aid for birdwatcher. 32-page booklet gives text about birds and songs, with illustration for each bird.

21829-5 Record, book, album. Monaural. $3.50

FADS AND FALLACIES IN THE NAME OF SCIENCE, Martin Gardner. Fair, witty appraisal of cranks and quacks of science: Atlantis, Lemuria, hollow earth, flat earth, Velikovsky, orgone energy, Dianetics, flying saucers, Bridey Murphy, food fads, medical fads, perpetual motion, etc. Formerly "In the Name of Science." x + 363pp.

20394-8 Paperbound $3.00

HOAXES, Curtis D. MacDougall. Exhaustive, unbelievably rich account of great hoaxes: Locke's moon hoax, Shakespearean forgeries, sea serpents, Loch Ness monster, Cardiff giant, John Wilkes Booth's mummy, Disumbrationist school of art, dozens more; also journalism, psychology of hoaxing. 54 illustrations. xi + 338pp.

20465-0 Paperbound $3.50

How to Know the Wild Flowers, Mrs. William Starr Dana. This is the classical book of American wildflowers (of the Eastern and Central United States), used by hundreds of thousands. Covers over 500 species, arranged in extremely easy to use color and season groups. Full descriptions, much plant lore. This Dover edition is the fullest ever compiled, with tables of nomenclature changes. 174 full-page plates by M. Satterlee. xii + 418pp.   20332-8 Paperbound $3.00

Our Plant Friends and Foes, William Atherton DuPuy. History, economic importance, essential botanical information and peculiarities of 25 common forms of plant life are provided in this book in an entertaining and charming style. Covers food plants (potatoes, apples, beans, wheat, almonds, bananas, etc.), flowers (lily, tulip, etc.), trees (pine, oak, elm, etc.), weeds, poisonous mushrooms and vines, gourds, citrus fruits, cotton, the cactus family, and much more. 108 illustrations. xiv + 290pp.   22272-1 Paperbound $2.50

How to Know the Ferns, Frances T. Parsons. Classic survey of Eastern and Central ferns, arranged according to clear, simple identification key. Excellent introduction to greatly neglected nature area. 57 illustrations and 42 plates. xvi + 215pp.   20740-4 Paperbound $2.00

Manual of the Trees of North America, Charles S. Sargent. America's foremost dendrologist provides the definitive coverage of North American trees and tree-like shrubs. 717 species fully described and illustrated: exact distribution, down to township; full botanical description; economic importance; description of subspecies and races; habitat, growth data; similar material. Necessary to every serious student of tree-life. Nomenclature revised to present. Over 100 locating keys. 783 illustrations. lii + 934pp. 20277-1, 20278-X Two volumes, Paperbound $7.00

Our Northern Shrubs, Harriet L. Keeler. Fine non-technical reference work identifying more than 225 important shrubs of Eastern and Central United States and Canada. Full text covering botanical description, habitat, plant lore, is paralleled with 205 full-page photographs of flowering or fruiting plants. Nomenclature revised by Edward G. Voss. One of few works concerned with shrubs. 205 plates, 35 drawings. xxviii + 521pp.   21989-5 Paperbound $3.75

The Mushroom Handbook, Louis C. C. Krieger. Still the best popular handbook: full descriptions of 259 species, cross references to another 200. Extremely thorough text enables you to identify, know all about any mushroom you are likely to meet in eastern and central U. S. A.: habitat, luminescence, poisonous qualities, use, folklore, etc. 32 color plates show over 50 mushrooms, also 126 other illustrations. Finding keys. vii + 560pp.   21861-9 Paperbound $4.50

Handbook of Birds of Eastern North America, Frank M. Chapman. Still much the best single-volume guide to the birds of Eastern and Central United States. Very full coverage of 675 species, with descriptions, life habits, distribution, similar data. All descriptions keyed to two-page color chart. With this single volume the average birdwatcher needs no other books. 1931 revised edition. 195 illustrations. xxxvi + 581pp.   21489-3 Paperbound $5.00

A HISTORY OF COSTUME, Carl Köhler. Definitive history, based on surviving pieces of clothing primarily, and paintings, statues, etc. secondarily. Highly readable text, supplemented by 594 illustrations of costumes of the ancient Mediterranean peoples, Greece and Rome, the Teutonic prehistoric period; costumes of the Middle Ages, Renaissance, Baroque, 18th and 19th centuries. Clear, measured patterns are provided for many clothing articles. Approach is practical throughout. Enlarged by Emma von Sichart. 464pp. 21030-8 Paperbound $3.50

ORIENTAL RUGS, ANTIQUE AND MODERN, Walter A. Hawley. A complete and authoritative treatise on the Oriental rug—where they are made, by whom and how, designs and symbols, characteristics in detail of the six major groups, how to distinguish them and how to buy them. Detailed technical data is provided on periods, weaves, warps, wefts, textures, sides, ends and knots, although no technical background is required for an understanding. 11 color plates, 80 halftones, 4 maps. vi + 320pp. 6⅛ x 9⅛. 22366-3 Paperbound $5.00

TEN BOOKS ON ARCHITECTURE, Vitruvius. By any standards the most important book on architecture ever written. Early Roman discussion of aesthetics of building, construction methods, orders, sites, and every other aspect of architecture has inspired, instructed architecture for about 2,000 years. Stands behind Palladio, Michelangelo, Bramante, Wren, countless others. Definitive Morris H. Morgan translation. 68 illustrations. xii + 331pp. 20645-9 Paperbound $3.00

THE FOUR BOOKS OF ARCHITECTURE, Andrea Palladio. Translated into every major Western European language in the two centuries following its publication in 1570, this has been one of the most influential books in the history of architecture. Complete reprint of the 1738 Isaac Ware edition. New introduction by Adolf Placzek, Columbia Univ. 216 plates. xxii + 110pp. of text. 9½ x 12¾. 21308-0 Clothbound $12.50

STICKS AND STONES: A STUDY OF AMERICAN ARCHITECTURE AND CIVILIZATION, Lewis Mumford.One of the great classics of American cultural history. American architecture from the medieval-inspired earliest forms to the early 20th century; evolution of structure and style, and reciprocal influences on environment. 21 photographic illustrations. 238pp. 20202-X Paperbound $2.00

THE AMERICAN BUILDER'S COMPANION, Asher Benjamin. The most widely used early 19th century architectural style and source book, for colonial up into Greek Revival periods. Extensive development of geometry of carpentering, construction of sashes, frames, doors, stairs; plans and elevations of domestic and other buildings. Hundreds of thousands of houses were built according to this book, now invaluable to historians, architects, restorers, etc. 1827 edition. 59 plates. 114pp. 7⅞ x 10¾ 22236-5 Paperbound $4.00

DUTCH HOUSES IN THE HUDSON VALLEY BEFORE 1776, Helen Wilkinson Reynolds. The standard survey of the Dutch colonial house and outbuildings, with constructional features, decoration, and local history associated with individual homesteads. Introduction by Franklin D. Roosevelt. Map. 150 illustrations. 469pp. 6⅝ x 9¼. 21469-9 Paperbound $5.00

MATHEMATICAL PUZZLES FOR BEGINNERS AND ENTHUSIASTS, Geoffrey Mott-Smith. 189 puzzles from easy to difficult—involving arithmetic, logic, algebra, properties of digits, probability, etc.—for enjoyment and mental stimulus. Explanation of mathematical principles behind the puzzles. 135 illustrations. viii + 248pp.
20198-8 Paperbound $2.00

PAPER FOLDING FOR BEGINNERS, William D. Murray and Francis J. Rigney. Easiest book on the market, clearest instructions on making interesting, beautiful origami. Sail boats, cups, roosters, frogs that move legs, bonbon boxes, standing birds, etc. 40 projects; more than 275 diagrams and photographs. 94pp.
20713-7 Paperbound $1.00

TRICKS AND GAMES ON THE POOL TABLE, Fred Herrmann. 79 tricks and games— some solitaires, some for two or more players, some competitive games—to entertain you between formal games. Mystifying shots and throws, unusual caroms, tricks involving such props as cork, coins, a hat, etc. Formerly *Fun on the Pool Table*. 77 figures. 95pp.
21814-7 Paperbound $1.25

HAND SHADOWS TO BE THROWN UPON THE WALL: A SERIES OF NOVEL AND AMUSING FIGURES FORMED BY THE HAND, Henry Bursill. Delightful picturebook from great-grandfather's day shows how to make 18 different hand shadows: a bird that flies, duck that quacks, dog that wags his tail, camel, goose, deer, boy, turtle, etc. Only book of its sort. vi + 33pp. 6½ x 9¼. 21779-5 Paperbound $1.00

WHITTLING AND WOODCARVING, E. J. Tangerman. 18th printing of best book on market. "If you can cut a potato you can carve" toys and puzzles, chains, chessmen, caricatures, masks, frames, woodcut blocks, surface patterns, much more. Information on tools, woods, techniques. Also goes into serious wood sculpture from Middle Ages to present, East and West. 464 photos, figures. x + 293pp.
20965-2 Paperbound $2.50

HISTORY OF PHILOSOPHY, Julián Marías. Possibly the clearest, most easily followed, best planned, most useful one-volume history of philosophy on the market; neither skimpy nor overfull. Full details on system of every major philosopher and dozens of less important thinkers from pre-Socratics up to Existentialism and later. Strong on many European figures usually omitted. Has gone through dozens of editions in Europe. 1966 edition, translated by Stanley Appelbaum and Clarence Strowbridge. xviii + 505pp. 21739-6 Paperbound $3.50

YOGA: A SCIENTIFIC EVALUATION, Kovoor T. Behanan. Scientific but non-technical study of physiological results of yoga exercises; done under auspices of Yale U. Relations to Indian thought, to psychoanalysis, etc. 16 photos. xxiii + 270pp.
20505-3 Paperbound $2.50

Prices subject to change without notice.
Available at your book dealer or write for free catalogue to Dept. GI, Dover Publications, Inc., 180 Varick St., N. Y., N. Y. 10014. Dover publishes more than 150 books each year on science, elementary and advanced mathematics, biology, music, art, literary history, social sciences and other areas.